Strategies for Success: Math Problem Solving, Grade 6 OT114 / 325NA   ISBN-13: 978-1-60161-939-6
Cover Design: Bill Smith Group   **Cover Illustration:** Valeria Petrone/Morgan Gaynin

Triumph Learning® 136 Madison Avenue, 7th Floor, New York, NY 10016 © 2011 Triumph Learning, LLC. Options is an imprint of Triumph Learning®. All rights reserved. No part of this publication may be reproduced in whole or in part, stored in a retrieval system, or transmitted in any form or by any means, electronic, mechanical, photocopying, recording or otherwise, without written permission from the publisher.

Printed in the United States of America.   10 9 8 7 6 5 4 3

Table of Contents

Problem-Solving Toolkit 5

Unit 1 Problem Solving Using Whole Numbers and Decimals 22

Unit Theme: Destinations

Lesson 1 **Draw a Diagram** 24
Math Focus: Operations with Whole Numbers

Lesson 2 **Work Backward** 30
Math Focus: Add and Subtract Decimals

Lesson 3 **Make a Table** 36
Math Focus: Multiply Decimals

Lesson 4 **Guess, Check, and Revise** 42
Math Focus: Divide Decimals

Review 48
Team Project: Stock a Store

Unit 2 Problem Solving Using Fractions 52

Unit Theme: Structures S.T.E.M.

Lesson 5 **Use Logical Reasoning** 54
Math Focus: Number and Fraction Concepts

Lesson 6 **Look for a Pattern** 60
Math Focus: Add and Subtract Fractions

Lesson 7 **Write an Equation** 66
Math Focus: Multiply Fractions

Lesson 8 **Solve a Simpler Problem** 72
Math Focus: Divide Fractions

Review 78
Team Project: Design a Roof

Unit 3 Problem Solving Using Algebra 82

Unit Theme: Explorations

Lesson 9 Work Backward 84
Math Focus: Integers

Lesson 10 Look for a Pattern 90
Math Focus: Variables and Expressions

Lesson 11 Write an Equation 96
Math Focus: Solve 1-Step Equations

Lesson 12 Solve a Simpler Problem 102
Math Focus: Solve 2-Step Equations

Review 108
Team Project: Plan an Adventure

Unit 4 Problem Solving Using Ratio, Proportion, Percent, and Probability 112

Unit Theme: At the Mall

Lesson 13 Make a Table 114
Math Focus: Rates and Ratios

Lesson 14 Use Logical Reasoning 120
Math Focus: Equivalent Ratios

Lesson 15 Write an Equation 126
Math Focus: Percent Concepts and Applications

Lesson 16 Make an Organized List 132
Math Focus: Probability

Review 138
Team Project: Make a Mall

Unit 5 — Problem Solving Using Geometry and Measurement 142

Unit Theme: Theme Park

Lesson 17 Guess, Check, and Revise................................ 144
Math Focus: Perimeter and Circumference

Lesson 18 Write an Equation....................................... 150
Math Focus: Area and Surface Area

Lesson 19 Draw a Diagram.. 156
Math Focus: Similar and Congruent Figures

Lesson 20 Write an Equation....................................... 162
Math Focus: Volume and Capacity

Review.. 168
Team Project: **Design a Park**

Unit 6 — Problem Solving Using Data and Graphing.................... 172

Unit Theme: The Great Outdoors

Lesson 21 Use Logical Reasoning................................... 174
Math Focus: Interpret Data

Lesson 22 Make a Graph.. 180
Math Focus: Circle Graphs and Bar Graphs

Lesson 23 Make an Organized List.................................. 186
Math Focus: Frequency Tables

Lesson 24 Make a Graph.. 192
Math Focus: Line Graphs

Review.. 198
Team Project: **Survey Your Class**

Glossary 202

Problem-Solving Toolkit

How to Solve Word Problems 6

Problem-Solving Checklist 7

Strategies

Look for a Pattern • Guess, Check, and Revise 8

Draw a Diagram • Write an Equation 9

Make a Table • Make an Organized List 10

Use Logical Reasoning • Make a Graph 11

Work Backward • Solve a Simpler Problem 12

Skills

Check for Reasonable Answers • Decide If an Estimate
or Exact Answer Is Needed 13

Decide What Information Is Unnecessary for Solving •
Find Hidden Information 14

Use Multiple Steps to Solve • Interpret Answers 15

Choose Strategies • Choose Operations 16

Solve Two-Question Problems • Formulate Questions 17

How to Read Word Problems 18

How to Solve Word Problems

Have you ever tried to get somewhere without clear directions? How did you find a way to get there? You can use the same kind of thinking when solving word problems.

- ☐ **Read the Problem** Read carefully to be sure you understand the problem and what it is asking. Try to visualize what is going on and what is being asked.

- ☐ **Search for Information** Look through all the words and all the numbers to see what information is given. Study any charts, graphs, and pictures. Anything that might help you solve the problem is important.

- ☐ **Decide What to Do** Think about the problem. If you are not sure how to solve it right away, ask yourself if you have solved a problem like this before. Think about all the problem-solving strategies you know. Choose one that you think will work.

- ☐ **Use Your Ideas** Start to carry out your plan. Try your strategy. Think about what you are doing. Once in a while, ask yourself, *Am I on the right track?* If not, change what you are doing. There is always something else you can try.

- ☐ **Review Your Work** Keep thinking about the problem. Finding an answer does not mean you are done. You need to keep going until you are sure you solved the problem correctly.

You can use the Problem-Solving Checklist on page 7 to make sure you have followed these important steps.

Problem-Solving Checklist

☐ Read the Problem

- ☐ Read the problem all the way through to get an idea of what is happening.
- ☐ Use context clues to help you understand unfamiliar words.

Ask Yourself

- ☐ How can I restate the problem in my own words?

☐ Search for Information

- ☐ Reread the problem carefully with a pencil in your hand. Circle the important numbers and math words.

Ask Yourself

- ☐ What do I already know?
- ☐ What do I need to find out to answer the question the problem asks?
- ☐ Does the problem have any facts or information that are not needed?
- ☐ Does the problem have any hidden information?
- ☐ Have I solved a problem like this before? If so, what did I do?

☐ Decide What to Do

- ☐ Choose a strategy that you think can help you solve the problem.
- ☐ Choose the operations you will use.

Ask Yourself

- ☐ How can I use the information I have to solve the problem?
- ☐ Will this problem take more than one step to solve?
- ☐ What steps will I use?

☐ Use Your Ideas

- ☐ Try the strategy you chose to solve the problem.
- ☐ Do the necessary steps.
- ☐ Write a complete statement of the answer.

Ask Yourself

- ☐ Do I need any tools such as a ruler or graph paper?
- ☐ Would an estimate of the answer help?
- ☐ Is my strategy working?

☐ Review Your Work

- ☐ Reread the problem.
- ☐ Check your computations, diagrams, and units.

Ask Yourself

- ☐ Is my answer reasonable? Does it make sense?
- ☐ Did I answer the questions the problem asks?

Problem-Solving Strategies

Look for a Pattern

Have you ever noticed that words in a song can sometimes be easier to remember than vocabulary terms? This is because a melody has a pattern, and vocabulary terms usually do not.

Patterns not only help us remember things, but they can help us solve math problems too.

Suppose you want to know the total number of boxes needed to create a pyramid-shaped display with 6 layers.

Do you see how you can solve the problem by using the pattern 1, 2, 3, 4, and so on?

> Be careful when you use patterns. What might look like a pattern at the start may not continue.

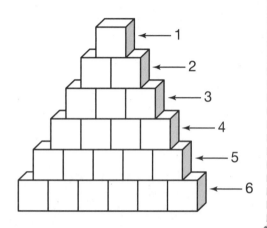

Guess, Check, and Revise

Have you ever forgotten which key unlocks your door? You try the keys one by one until the lock opens. You might be lucky and find the right key first, or it might take longer.

In word problems, you can make guessing work faster. First, guess an answer and see if it is correct. When a guess does not work, think about how to change it to find the answer you want.

The combination to a lock consists of two 2-digit numbers. The sum of the numbers is 40, and their product is 336. What are the two 2-digit numbers in the combination?

Could the numbers be 30 and 10? No, because 30×10 is 300, which is too low. How about 20 and 20? No, 20×20 is 400, which is too high. Keep revising your guesses until you get two numbers that match the clues.

> When you use Guess, Check, and Revise, use what you learn from one guess to help you make your next guess.

Draw a Diagram

During the Revolutionary War, George Washington had his soldiers build and float a heavy link chain across the Hudson River to block British ships. The chain was about 1,500 feet long. Each link weighed 114 pounds. Can you picture this heavy chain?

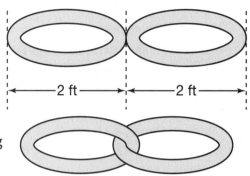

Picturing a situation helps you understand what is going on. Drawing a diagram can help you understand more.

Each link in the chain was 2 feet long and made of metal $2\frac{1}{2}$ inches thick. How long would a chain of 10 of these links be?

Drawing a diagram helps you see how the links overlap. You can see that the length of a 10-link chain is not 10 × 2 feet. Can you figure out the actual length?

A diagram can help you see how all the parts of a problem are related. You can draw any kind of diagram that helps you picture the situation.

Write an Equation

An equation is like a recipe. It tells you what steps to follow to get the result you want.

Suppose you want to find the distance across the inside of a link in the chain. You can write an equation in words to show how to calculate that distance.

inside distance = entire length − (2 × thickness)

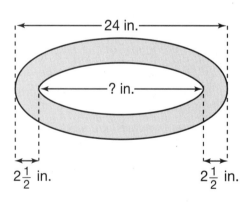

You can also write the equation with symbols. Use D to stand for the inside distance.

$D = 24$ in. $- (2 \times 2\frac{1}{2}$ in.$)$

To find the answer, solve the equation.

Writing an equation in words first can help you describe the situation. Then use numbers and symbols to write the same equation.

Problem-Solving Toolkit

Problem-Solving Strategies

Make a Table

Choose a note in Box A without telling a partner which one you chose. Next, describe to your partner without using pictures or gestures where your note is located in the box.

Now try this activity again with Box B.

Describing your note is easier when the notes are arranged in rows and columns. Organizing information in tables can make solving problems easier, too.

Ari has saved $38 to buy a guitar that costs $129. If he saves $21 each week, in how many weeks will he have enough money to buy the guitar?

You can use a table to keep track of Ari's savings.

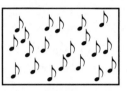
Box A

Box B

Week	Now	1	2	3	4
Savings	$38	$59	$80	$101	$122

When you have two or more kinds of data, you can use a table to organize the information.

Make an Organized List

When the director selected the cast for her play, she chose from among 5 actors who tried out for 3 roles. How many different groups of 3 actors might be chosen from a group of 5?

To answer the question, you can list all the possible groups of 3. You can use letters in your list to stand for each actor. Keeping track of the letters you list will help you include every possible group once and only once.

```
ABC   ABD   ABE
ACD   ACE
ADE

BCD   BCE
BDE

CDE
```

Before making a list, think about how you can make it in an organized way. Decide what you will put first in your list.

Use Logical Reasoning

What day was it that you had a burrito for lunch in the school cafeteria? It could not have been Wednesday, because you always bring lunch on Wednesday. It could not be Friday, either, because spaghetti is always served on Friday.

You can keep eliminating days of the week until there is only one day left. This kind of logical reasoning can also be used to solve math problems.

The mystery number could be any digit from 0 to 9. When you spell it, there is no letter *e* and no letter *o*. Which digits do not work? Which digit must it be?

> You can use a list or a table to help you keep track of the choices you eliminate by using logical reasoning.

zero	four	eight
one	five	nine
two	six	
three	seven	

Make a Graph

Suppose two baby pandas are born on the same day.

Baby A has a mass of 2.3 kilograms at 40 days old, 2.9 kilograms at 50 days old, 3.1 kilograms at 60 days old, 3.7 kilograms at 70 days old, and 4.3 kilograms at 80 days old.

Baby B has a mass of 1.9 kilograms at 40 days old, 2.5 kilograms at 50 days old, 3.5 kilograms at 60 days old, 4.0 kilograms at 70 days old, and 4.1 kilograms at 80 days old.

You want to know if there are any days that Baby B weighed more than Baby A.

A graph makes it easier to solve the problem.

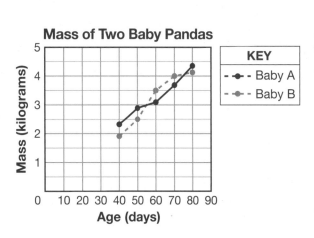

> When you have two kinds of data and want to know how they are related, you can make a graph to see the relationship.

Problem-Solving Toolkit 11

Problem-Solving Strategies

Work Backward

Matt works at a pet store every Saturday beginning at 1 P.M. It takes him 15 minutes to get there from his house. He needs a half hour to eat lunch before he goes. Also, he has an hour of chores to do before he leaves home. What is the latest time Matt can start his chores and still make it to the pet store on time?

You can work backward to find out.

- To be at the pet store by 1:00, Matt has to leave the house by 12:45.
- To finish eating lunch by 12:45, Matt needs to start eating by 12:15.
- So Matt has to start his chores at least an hour before 12:15.

What time would that be?

If you know an end result, and you know what happens at each step toward that end result, you can work backward to find the starting point.

Solve a Simpler Problem

Bella started pulling weeds at 10:15 A.M. By 10:40 she had pulled 36 weeds. At that rate, how many weeds would she have pulled when she stops at 11:30 A.M.?

This problem may seem difficult, but you can make it simpler. Try changing the numbers to make them easier to work with.

Bella started pulling weeds at **10:00 A.M.** By **10:15** she had pulled **50** weeds. At that rate, how many weeds would she have pulled when she stops at **11:00 A.M.?**

How do you solve the simpler problem? You can use the same steps to solve the original problem.

If a problem seems complicated, try to break it down into a simpler problem. You can make the numbers as easy as you want them to be.

Problem-Solving Skills

Check for Reasonable Answers

There is an old joke about an airplane pilot who says, "I flew $\frac{3}{4}$ of the way and then realized there was not enough fuel, so I went back to get more." Common sense tells us the pilot's thinking is not reasonable. Common sense is an important tool when you are solving math problems.

Suppose you help your parents figure out how many miles their car travels on 1 gallon of gas. Your family drove 289 miles and used 6.3 gallons. On your calculator, you divide 289 by 6.3, and you get 458.73.

You know that answer cannot be right. It does not make sense.

Do you see why? What might have happened when you calculated?

> Whenever you solve a problem, look back to see if your answer is reasonable. Checking for reasonableness helps you spot and fix any mistakes.

Decide If an Estimate or Exact Answer Is Needed

You work at the library after school. So far this week you worked $1\frac{3}{4}$ hours on Monday and the same amount of time on Tuesday. You also put in $1\frac{1}{2}$ hours on Wednesday and again on Thursday. You are not allowed to work more than 10 hours a week. You want to find out if you have worked over 10 hours so far.

Hours so far = $1\frac{3}{4} + 1\frac{3}{4} + 1\frac{1}{2} + 1\frac{1}{2}$

Without calculating the exact answer, can you tell if you worked more than 10 hours?

> Be sure that you need an exact answer before you start to calculate. Sometimes, an estimate is all you need to find an answer.

Problem-Solving Toolkit

Problem-Solving Skills

Decide What Information Is Unnecessary for Solving

You are doing some research about stars in the galaxy for a science project.

The Web site you are using for your research has more information than you need. But you know how to ignore the unnecessary information and find what you need. This kind of skill also helps when you solve math problems.

Solving a math problem can be like looking at a Web site filled with information. Knowing what information you need is helpful so that you can ignore information you do not need.

Find Hidden Information

What can you do? The article you want is on page 37, but there is no page 37. Oh, there it is! The page between 36 and 38 has no page number. So page 37 was there all along, hiding right in the middle of all the numbered pages.

Many math problems also have information hidden from plain view. Diagrams, tables, and graphs may contain data you have to search for.

Sometimes, you have to supply information yourself. You may need to use the fact that there are 12 months in a year or that there are 1,000 meters in a kilometer.

When the information you need is not clearly stated in a problem, read the problem again carefully and examine any diagrams, tables, or graphs.

Use Multiple Steps to Solve

Before you cross the stream, you need to walk to the bridge. Most things, including math problems, involve more than one step.

At the park, a giant balloon usually costs $7.50. How much will it cost on sale?

For a first step, you can find $\frac{1}{3}$ of $7.50 so you know how much will be taken off the price.

What will be your next step to find the answer?

Reread a problem to decide how many steps you need to use.

Interpret Answers

Does this picture look like it is moving? Sometimes our brains interpret pictures in certain ways, whether we want them to or not.

Interpreting has a role in solving math problems. You may have made a calculation, but your calculation might not mean that you have answered the question.

The school is forming teams for its Volleyball Challenge. Each team must have exactly 6 players. So far, 38 players have signed up. How many more players need to sign up so that all the teams have 6 players and there are no players without teams?

You can divide. $38 \div 6 = 6\ R2$

Can each team have 6 R2 players? You have to interpret 6 R2 so that your answer to the question in the problem makes sense.

Think about the problem situation to determine if you need to interpret an answer.

Problem-Solving Skills

Choose Strategies

Should I move this game marker, or should I play it safe and keep one in the home square?

With some games you can increase your chance of winning by choosing a strategy. Using an effective strategy can also increase your chance of success when solving a math problem.

In a game, you are trying to decide which of two paths to take. Which path is shorter?

- To decide, you might use the strategy *Draw a Diagram*.
- You might also try the strategy *Use Logical Reasoning*.
- You might use some other strategy.

When you are solving math problems, you can choose the strategy that works best for you. If one strategy does not seem to be working, then try a different strategy.

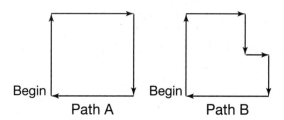

Begin — Path A   Begin — Path B

Choose Operations

In the game *Explorland*, you walk around the whole region you will explore.

Suppose you walk 50 yards north, 20 yards east, 50 yards south, and 20 yards west to get back to your starting point. What is the area of the rectangular region that you walked around?

Math problems like this one can be simple and direct. One operation may be all you need.

Could you find the area with a single step? Would you add, subtract, multiply, or divide?

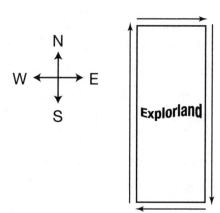

When you solve a problem, think first before calculating. You may be able to solve some problems without calculating at all.

Solve Two-Question Problems

> When a math problem asks more than one question, take the time to think about each one and answer it.

When you order a meal, you may need to answer two or more questions. You may be asked which kind of sauce or which kind of drink you would like. When you solve a math problem, you may get two questions, too.

Suppose you order one salad, a chicken sandwich, and two cups of juice. How much will your entire order cost? If you pay with a $20 bill, how much change should you get?

Can you answer both questions?

MENU
Salad	$2.50
Sandwiches	$5.00 each
Chicken	
Tuna	
Roast Beef	
Veggie	
Juice	$2.00 each
Orange	
Apple	

Formulate Questions

Have you ever gazed at a machine and wondered how it works? Sometimes, the inside reveals a mystery. The closer you look, the more questions you may have.

Mathematicians formulate questions. They look at shapes and numbers and wonder how they work. What they learn depends on the kind of questions they ask, and answers lead to more questions.

As you look at the gears, you might wonder which way each gear turns.

What other questions might you ask about the gears?

> Formulating questions about a math problem can help you understand it better.

How to Read Word Problems

Word problems usually do three things.

- They give you some information.
- They tell you how parts of that information are related.
- They ask a question or set a goal.

Read to Understand

When you read a word problem, you may see words or symbols that are new to you. Here are some things you can do.

- You can **use context clues** to help you understand.
- You can **look up** the word, fact, or symbol.
- You can use **root words** to figure out the meaning.

You need to read a word problem carefully to understand it. Sometimes, reading again with a special purpose will help.

Read the problem. Then write the meaning of each underlined word.

> The Lees have <u>quadruplets</u>. Mrs. Lee is waiting in line to buy <u>berets</u> for the 4 girls. She will get a <u>discount</u> of 25%. The regular price is $9.79 each. How much will Mrs. Lee pay in all, not including tax?

1. I can use context clues to figure out the meaning of <u>quadruplets</u>.

   Meaning _____
   _____

2. I can look up the word <u>beret</u>.

   Meaning _____
   _____

3. I can use a root word to understand <u>discount</u>.

   Meaning _____
   _____

Sometimes, a word in a problem can have more than one meaning. Compare the math meaning of *line* to its everyday meaning.

Everyday Meaning _____

Math Meaning _____

Look for Information

Read a problem once to be sure you understand what it is about. Then read it again to identify numbers and words you need.

▶ A problem may give you all the information you need, and nothing else.

   A map's scale is (1 inch = 2 miles.) A trail is about (2.5 inches) long on the map. How many miles is the trail?

▶ A problem may have all the information you need and some extra details.

   You have a 3-foot-square map with a scale of (1 inch = 2 miles.) The trail you want to walk is (about 2.5 inches) long on the map. How many miles will you walk?

   You do not need this to solve the problem.

▶ A problem may have only some of the details you need. You must find a way to get the rest of the information.

   Your map of Bell Park has a scale of 1 inch = 0.1 mile. You want to (walk from the lake to the meadow.) How many miles will you walk?

   You need to measure the distance on the map.

▶ Some data you need may be in tables, graphs, or diagrams.

Read each problem. Study the information to decide how it can help you solve the problem. Then write your answer.

1. Sam says that triangle A is congruent to triangle B. Is he correct? Explain.

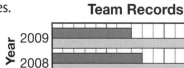

Information I can get only from the diagram: _____

_____

2. Each year the Sluggers and the Runners play 20 games. In which year did the Sluggers come closer to winning more games than the Runners?

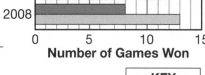

Data I can get only from the graph: _____

_____

Problem-Solving Toolkit 19

How to Read Word Problems (continued)

 Mark the Text

Marking information you need in the text as you read can help you organize your thinking.

- You can circle numbers and words.
- You can cross out information you do not need.
- You can underline the question the problem asks.
- You can also mark something you do not understand or need to look up.

> Blake is planning a mural for West Park. He has ~~6 weeks~~ to paint it. Before Blake starts the mural, he makes a sketch. The sketch is (25 inches wide and 15 inches tall. The mural will be 20 feet wide.) How tall will the mural be?

The question does not ask how long Blake has to paint the mural, so I do not need this detail.

I need the dimensions of the sketch and the width of the mural.

Here is the question I need to answer.

Mark the text and tell why each mark is important.

> Lucy and her friends went shopping for school. Lucy had $20 to spend. Lucy bought a package of pencils for $1.99, a package of pens for $2.99, and a notebook for $5.49. The sales tax is 6% in her state. How much did Lucy spend on school supplies?

I underlined _____

because _____

I crossed out _____

because _____

I circled _____

because _____

Decide What to Do

You can look at a problem to be sure you know what question you need to answer.

▶ Sometimes a word problem clearly asks a question that you can answer by computing.

> The Panthers won 12 out of 20 games. <u>What percent of their games did they lose?</u>

Describe What steps will you take to solve the problem?

_____

_____

▶ Sometimes you need to compute, but the result of your computation is not the answer to the problem.

> The Panthers won 12 out of 20 games. <u>Did they lose more than 40% of their games?</u>

Compare How is this problem like the first problem on this page?

_____

_____

▶ Sometimes you do not need to compute at all. You can find the answer using a different method.

> The Panthers won 12 out of 20 games. <u>Did they lose more than 50% or less than 50% of their games?</u>

Consider Do you need to compute to answer the question? Explain.

_____

_____

Problem-Solving Toolkit   21

UNIT 1: Problem Solving Using Whole Numbers and Decimals

Unit Theme:
Destinations

How do you reach your destinations? What do you do when you get there? Maybe you go to a lake to paddle in a canoe. Some people take planes or trains to visit different cities. In this unit, you will see the many ways math is used at your destinations and along the way.

Math to Know

In this unit, you will apply these math skills:

- Compute with whole numbers
- Add and subtract decimals
- Multiply and divide decimals

Problem-Solving Strategies

- Draw a Diagram
- Work Backward
- Make a Table
- Guess, Check, and Revise

Link to the Theme

Write another paragraph about the basketball camps. Include some of the words and numbers from the table at the right.

Lynn wants to go to basketball camp this summer. There are three camps from which to choose. She puts the prices in a table to show her parents.

Camp Name	Price (per week)
Champs Basketball	$279.50
Hoops Basketball	$315.00
Super Sports	$299.99

_____

_____

_____

_____

_____

_____

22

Use Math Language

Review Vocabulary

The list below shows vocabulary terms in this unit. Knowing the meaning of these terms will help you understand the problems.

bar diagram	depth	multiply	Venn diagram
decimal	difference	product	volume

Vocabulary Activity Multiple-Meaning Words

Some terms have more than one meaning. Use terms from the list above to complete the following sentences.

1. Lenny waited until the new _____ went on sale before buying it.

2. The _____ is the answer I get when I multiply.

3. Dana used the remote control to lower the _____ on the television.

4. _____ is the measure of the amount of space enclosed by a solid figure.

Graphic Organizer Word Map

Complete the graphic organizer.

- Write a definition of the term *decimal*.
- Use the term in a sentence.
- Show three examples of the term.
- Show an example of a number that is *not* a decimal.

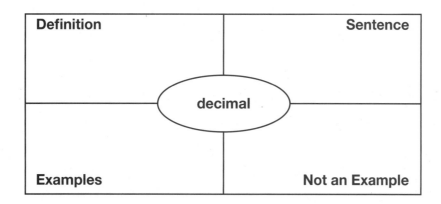

23

Strategy Focus
Draw a Diagram

MATH FOCUS: Operations with Whole Numbers

Learn

▣ Read the Problem

> Mr. Darrell surveyed his students to find out how many had traveled by boat or by train. Eleven students had traveled by boat and 16 students had traveled by train. Four students had traveled by *both* boat and train. There are 31 students in the class. How many students had never traveled by boat or by train?

Reread Ask yourself these questions as you read.

- What is this problem about?

  _____

- What kinds of information are given?

  _____

- What do you need to find out?

  _____

▣ Search for Information

Mark the Text

Read the problem again. Circle words and numbers that will help you solve the problem.

Record List the details from the problem.

Number of students who had traveled by boat: _____

Number of students who had traveled by train: _____

Number of students who had traveled by *both* boat and train: _____

Number of students in the class: _____

You can use this information to choose a problem-solving strategy.

Decide What to Do

The problem describes two groups of students. Students might belong to one, both, or neither group. So the groups may overlap.

Ask How can I find the number of students who had never traveled by boat or by train?

- I can use the strategy *Draw a Diagram*. I can draw a Venn diagram.

Use Your Ideas

Step 1 Draw a Venn diagram. The section where the circles overlap shows "Both Boat and Train." Write 4 in this section.

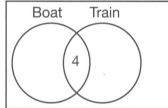

When a problem has groups that overlap, a Venn diagram can be helpful.

Step 2 Find the numbers of students for the other sections.

Only by Boat: altogether, 11 traveled by boat; 4 traveled by both boat and train. Subtract. Write the answer in the "Boat" section.

11 − 4 = _____

Only by Train: altogether, 16 traveled by train; 4 traveled by both train and boat. Subtract. Write the answer in the "Train" section.

16 − 4 = _____

Step 3 Find the number of students that never traveled by boat or train.

Find the total number of students who have traveled by boat or train. Subtract that sum from the total number of students in the class. Write the answer outside the circles but inside the frame.

31 − 23 = _____

So _____ students had never traveled by boat or by train.

Review Your Work

Check that the numbers in the Venn diagram match the problem.

Describe How did the Venn diagram help you solve this problem?

_____

_____

Try

Solve the problem.

1) Ryan and Emma keep track of the number of miles they hike. Ryan compares his total with Emma's. Emma hiked 16 miles more than Ryan. Together, they hiked 250 miles. How far did Emma hike?

▪ Read the Problem and Search for Information

Reread the problem. Cross out information that is not needed.

▪ Decide What to Do and Use Your Ideas

You can use the strategy *Draw a Diagram*. A bar diagram can help you compare the distances Ryan and Emma hiked.

Step 1 Draw a bar diagram. The whole is the total distance Ryan and Emma hiked.

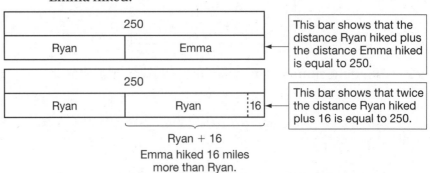

Ask Yourself

How can I show how the number of miles Ryan hiked relate to the number of miles Emma hiked?

Step 2 Find the distance Ryan hiked. Mark this on the diagram.

250 − 16 = _____

234 ÷ 2 = _____

Step 3 Find the distance Emma hiked.

117 + 16 = _____

So Emma hiked _____ miles.

▪ Review Your Work

Check that your answers match the information in the problem.

Identify In Step 2, what does the number 234 represent?

_____

_____

26 Unit 1 **Using Whole Numbers and Decimals**

Apply

Solve the problems.

(2) A shop rents out only bikes and rafts. On Saturday, 75 people rented from the shop. Sixty-eight people rented rafts. Thirty-five people rented both bikes and rafts. How many people rented bikes in all?

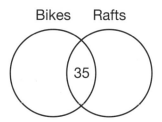

Ask Yourself
What does the number 35 in the middle section show?

Answer _____

Summarize Describe the three sections within the Venn diagram.

_____

_____

Hint The total number of people who rented rafts includes the number of people who rented both bikes and rafts.

(3) Kaitla is renting a canoe at a lake. She plans to canoe for 3 hours. The shop rents canoes by the hour or by the day. It costs $70 to rent a canoe for a full day. Kaitla figures out that it would cost $16 more to rent the canoe for a full day than to rent it for 3 hours. How much does it cost to rent a canoe for 1 hour?

Hint The canoe is being rented for 3 hours. You need to find the cost for 1 hour.

Complete the bar diagram.

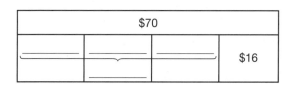

Ask Yourself
Is the cost for a full day $16 more than the cost per hour or more than the cost for all 3 hours?

Answer _____

Generalize How can a bar diagram be helpful in solving problems?

_____

_____

Lesson 1 **Strategy Focus: Draw a Diagram** 27

④ Paul is training for a race in which he will swim, bike, and run. He keeps track of his training for 100 days. Paul runs every day. He bikes a total of 82 days. On 21 of those days, he also swims. He does not bike or swim on 15 days. How many days did Paul swim?

Hint Start by filling in the number of days Paul both bikes and swims.

Ask Yourself
How many days in total does Paul bike or swim?

[Venn diagram with two overlapping circles labeled "Bike" and "Swim"]

Answer _____

Determine What information is not needed to solve the problem?

_____

_____

⑤ On Monday, Wednesday, and Friday, Pam took the train from home to her aunt's house and back. It is an 8-mile trip each way. On Tuesday and Thursday, she rode the train from home to science camp and back. Pam rode 124 miles on the train that week. How far is her ride to camp each way?

Hint Find how many 8-mile trips Pam made.

Ask Yourself
How should I divide the bar diagram?

124

Answer _____

Analyze Sid thinks that the ride to science camp is 76 miles. What mistake might Sid have made?

_____

_____

28   Unit 1   **Using Whole Numbers and Decimals**

Practice

Solve the problems. Show your work.

6 Mike's family is driving to New York City. The trip is 510 miles long. Forty-two miles before the halfway point of the trip, they stop for lunch. How much farther do they have to drive to get to New York City?

Answer _____

Conclude After you found the halfway point, did you add or subtract? Explain why.

_____
_____

7 Maya's school took a field trip to an amusement park. There were 238 students on the trip. A total of 145 students rode water rides. Of those, 53 students also rode bumper cars. Eight students did not ride any water rides or bumper cars. How many students rode bumper cars, but not water rides?

Answer _____

Develop What is another question you could ask about the problem?

_____
_____

Create

Look back at Problem 4. Write a new problem with a different number of days that Paul both swims and bikes. Solve your new problem.

Lesson 1 **Strategy Focus: Draw a Diagram** 29

Lesson 2

Strategy Focus
Work Backward

MATH FOCUS: Add and Subtract Decimals

Learn

Read the Problem

Mrs. Cora travels to craft fairs to sell her hand-crafted jewelry. Last Saturday, she sold some jewelry in the morning. On her break, she spent $24.86 of the money she earned to buy lunch and a hand knit sweater. Then she sold another $34.80 worth of jewelry in the afternoon. At the end of the day she had $50.00 left of the money that she earned by selling jewelry. How much money did Mrs. Cora earn selling jewelry in the morning?

Reread Visualize what is happening, and describe it in your own words.

- What happened at the craft fair in the morning?
  _____

- What happened on Mrs. Cora's break?
  _____

- What happened in the afternoon?
  _____

- What do you have to find out?
  _____

Mark the Text

Search for Information

As you read the problem again, underline details. Think about the sequence of Mrs. Cora's day.

Record What details do you know that will help you solve the problem?

On her break, Mrs. Cora spent _____ on lunch and a sweater.

In the afternoon, Mrs. Cora sold _____ worth of jewelry.

At the end of the day, Mrs. Cora had _____ .

Use this information to help you choose a strategy to find how much money Mrs. Cora earned selling jewelry in the morning.

30 Unit 1 Using Whole Numbers and Decimals

Decide What to Do

You know how much money Mrs. Cora had at the end of the day, how much she earned in the afternoon, and how much she spent on her break. You also know she earned some money in the morning.

Ask How can I find out how much money Mrs. Cora earned selling jewelry in the morning?

- I can use the strategy *Work Backward*.
- I know how much money Mrs. Cora had at the end of the day. I know how the amount of money changed in each part of her day.
- I can undo each step to find how much money she earned in the morning.

Use Your Ideas

Step 1 Write how the amount of money Mrs. Cora had changed at each part of her day.

Step 2 Work backward. Use inverse operations to undo each step.

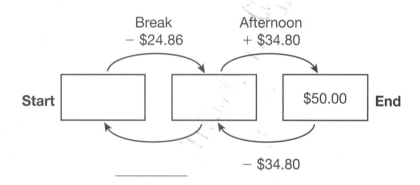

To undo adding $34.80, subtract $34.80.

Mrs. Cora earned _____ selling jewelry in the morning.

Review Your Work

You worked backward to solve this problem. Start with your answer and work forward through Mrs. Cora's day to check your work.

Identify When is the strategy *Work Backward* a good strategy to use?

_____

_____

Try

Solve the problem.

(1) Three friends are traveling to Seattle from Saint Louis. Kira got her plane ticket for $34.50 less than Dan's ticket. Elena paid $17.13 less than Kira. Elena's ticket cost $219.34. How much did Dan's ticket cost?

Read the Problem and Search for Information

Look for details in the problem that show the relationships between the ticket prices. Restate the relationships in your own words.

Decide What to Do and Use Your Ideas

You can use the strategy *Work Backward* to find the cost of Dan's ticket.

Step 1 Draw a diagram to show how the ticket prices relate to each other.

Step 2 Work backward. Use inverse operations to undo each step.

Ask Yourself

I know that Elena paid less than Kira. What operation should I use to find Kira's ticket price?

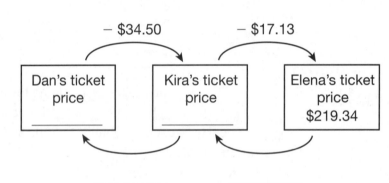

Dan's plane ticket cost _____.

Review Your Work

Did you use the correct number for the cost of each person's ticket?

Explain Will Dan's ticket cost more or less than Kira's? Than Elena's? How do you know?

_____

_____

32   Unit 1   **Using Whole Numbers and Decimals**

Apply

Solve the problems.

2) Lake A has the lowest rate for a four-hour canoe rental. Lake B charges $3.45 more than Lake A. Lake C charges $1.25 less than Lake B. It costs $26.00 for a four-hour rental at Lake C. How much does a four-hour canoe rental cost at Lake A?

Ask Yourself
What do the words *more* and *less* tell me about the operations I should use?

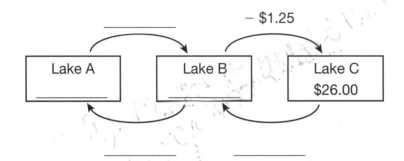

◀ **Hint** Order the lakes so that the costs that are related are next to each other.

Answer _____

Conclude What information is not necessary to solve the problem?

_____

_____

3) Jaime bought a used metal detector for $29.98. Using it, he found some money on Monday. On Tuesday, he found $12.13, but he lost four dimes on the way home. He figures out that if he finds $6.85 more, he will have found the amount of money equal to what he paid for the metal detector. How much money did Jaime find on Monday?

◀ **Hint** Make sure to think about whether each step adds to or subtracts from Jaime's total.

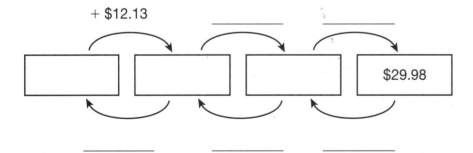

Ask Yourself
Why should the diagram end with $29.98?

Answer _____

Interpret What is another way you could solve this problem?

_____

_____

Lesson 2 **Strategy Focus: Work Backward** 33

Hint Make sure that you pay attention to which restaurants are being compared to each other.

 Ask Yourself
In what order should I show the restaurants in the diagram?

(4) Alex's favorite chowder is at Davy's Clam Shack. The chowder at Sand Dollar costs $1.19 less. Chowder costs $4.00 at Seaview. That is $0.89 more than at Sand Dollar. How much does the chowder cost at Davy's Clam Shack?

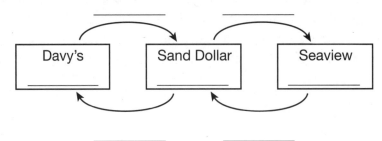

Answer _____

Determine How would your answer change if the chowder at Seaview cost $3.50 instead of $4.00?

_____

_____

(5) Mr. Ortiz is on a business trip and he is keeping track of his spending. He has two twenty-dollar bills in his wallet at the start of the day. He goes to the ATM and takes out some money. His lunch costs $11.30. His dinner costs $22.20. He has $36.50 left and does not spend any more money. How much money did Mr. Ortiz take out at the ATM?

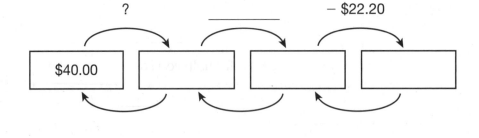

Hint Put all the details from the problem into the diagram.

 Ask Yourself
Which operations should I use?

Answer _____

Apply Do you need the information that Mr. Ortiz does not spend any more money after dinner to solve the problem? Why or why not?

_____

_____

Practice

Solve the problems. Show your work.

6 Mr. Lee is planning a train trip to Dallas, Chicago, and New York. His ticket to Dallas will be the cheapest. His ticket to Chicago costs $56.86 more than his ticket to Dallas. His New York ticket costs $176.10. That is $32.90 more than his Chicago ticket. What is the total cost of Mr. Lee's trip?

Answer _____

Infer Jay answered that the trip cost $86.34. What mistake might Jay have made?

_____

_____

7 Miko wants to go camping. Her father said that if they could buy a tent and a stove for less than $140, they could go. They bought a tent. Then they paid $53.99 for a stove. They still had $15.50 left of the $140. How much did the tent cost?

Answer _____

Discuss How does working backward help you solve this problem?

_____

_____

Create Look back at Problem 4. Write and solve a new problem about something you like to eat. Change at least two of the numbers in the original problem. Be sure your problem can be solved using the strategy *Work Backward*.

Lesson 2 **Strategy Focus: Work Backward** 35

Lesson 3

Strategy Focus
Make a Table

MATH FOCUS: Multiply Decimals

Learn

Read the Problem

Sandy is making a sandcastle with a rectangular base. She pushes seashells into the bottom border of her castle. 13 seashells fit along the front and another 13 fit along the back. There are 7 seashells along each of the sides. The ones on the front and back are 4.25 inches wide. The seashells on the sides are 3.5 inches wide. What is the total distance around Sandy's sandcastle?

Reread Ask yourself these questions as you read the problem.

• What is this problem about?
_____

• What do you need to find out?
_____

Search for Information

Mark important details and underline the question you will answer.

Record Write the details from the problem.

Each shell on the front and back is about _____ inches wide.

_____ seashells fit along the front of the castle.

_____ seashells fit along the back of the castle.

Each shell on the sides is about _____ inches wide.

_____ seashells fit along the left side of the castle.

_____ seashells fit along the right side of the castle.

The strategy you choose should help you organize this information.

36 Unit 1 **Using Whole Numbers and Decimals**

Decide What to Do

You know how many seashells are on each side. You know the width of each shell. You want to find a total length.

Ask How can I find the distance around Sandy's sandcastle?

- I can use the strategy *Make a Table*.
- I can multiply the number of shells on each side by the width of one shell to find the length of that side. Then I can add the lengths of the sides to find the total distance around the castle.

Use Your Ideas

Complete the table below as you follow these steps.

Step 1 Fill in the number of shells and the width of each shell.

Step 2 Multiply the number of shells by the width of the shell.

Step 3 Add to find the total distance around Sandy's castle.

Side	Number of Shells	Width of Shell (in.)	Distance (in.)
Front	13	4.25	55.25
Back			
Left			
Right			
		Total	

Tables help you organize information and help you keep track of your work.

The distance around Sandy's sandcastle is _____ inches.

Review Your Work

Check your multiplication by rounding the width of each shell to the nearest whole number.

Describe How did the table help you to solve the problem?

_____

_____

Try

Solve the problem.

1) Bob and Betty are snorkeling. They see some blue-striped grunt fish and a leatherback turtle. Bob guesses that it would take at least 8 grunt fish in a line to be as long as one turtle. Betty says 6 grunt fish would be enough. One grunt fish is about 24.5 centimeters long. The turtle is about 161.2 centimeters long. Who made the closer guess?

Mark the Text

Read the Problem and Search for Information

Locate the mathematical details in the problem. Review the information and think about how you will organize all the facts.

Decide What to Do and Use Your Ideas

Use the strategy *Make a Table* to collect all the facts from the problem.

Step 1 Put the details from the problem into the table and find the total length for each guess.

Ask Yourself

I know how many grunt fish are in a line, and how long each fish is. What operation should I use to find the total length of the line?

	Number of Fish	Length of One Fish (cm)	Total Length (cm)	Difference from Turtle's Length (cm)
Bob		24.5		
Betty				

Step 2 Subtract to find the difference between each guess and the length of the turtle.

So _____ made the closer guess.

Review Your Work

Use a whole number close to the length of the grunt fish to check your multiplication.

Conclude Steven said that the column for the length of one grunt fish is unnecessary. Do you agree? Explain.

_____

_____

Apply

Solve the problems.

2) Nala swims in a rectangular pool that is 15.24 meters long and 5.62 meters wide. She calls swimming back and forth across the width a "little lap." Swimming back and forth across the length is a "big lap." Nala swims 10 big laps and 5 little laps. How many meters does she swim?

	Number of Laps	Length for Each Lap (meters)	Distance Nala Swims (meters)
Big Laps	10	30.48	
		Total	

◀ **Hint** Set up your table with a row for each kind of lap.

Ask Yourself
What does 30.48 represent in the table?

Answer _____

Explain Why is it important to compute the total length for each kind of lap separately?

_____

_____

3) Dara, Phil, and Andy are digging rectangular holes. All their holes are 1 foot deep. Dara's hole is 6.5 feet long and 3.75 feet wide. Phil's is 5.5 feet long and 4.75 feet wide. Andy's is 7.5 feet long and 3.25 feet wide. Which hole has the greatest volume?

You know the depth of each hole is _____ .

Remember that volume is measured in _____ units.

Name	Length (ft)	Width (ft)	Depth (ft)	Volume (ft³)
Dara			1	
Phil				

◀ **Hint** Multiply length, width, and depth to find the volume of each hole.

Ask Yourself
Where do I place the decimal point when multiplying two decimal numbers?

Answer _____

Categorize What types of problems can the strategy *Make a Table* help you to solve?

_____

_____

Lesson 3 **Strategy Focus: Make a Table** 39

Ask Yourself

How many digits will there be to the right of the decimal point when I multiply 2.619 and 11.2?

Hint Round the cost to the nearest penny.

④ Cara is going to the beach. She bought 11.2 gallons of gas for $2.619 per gallon. Then she saw a station selling cheaper gas. If she had waited, she could have paid $2.539 per gallon. How much money would Cara have saved if she had waited?

Price per Gallon	Gallons	Cost
	Cara would have saved	

Answer _____

Identify How did you know what operation to use?

_____

_____

Ask Yourself

How can I organize the sizes and numbers of the shells?

Hint Be careful to match the correct length and number with each kind of shell.

⑤ Tina has shells along the front of her garden. Conch shells are 8.25 inches long. Sand dollars are 2.5 inches long. Scallop shells are 1.5 inches long. There are 10 conch shells, twice as many sand dollars as conch shells, and 4 times as many scallop shells as conch shells. How long is the front of the garden?

Kind of Shell	Length of One Shell (in.)	Number of Shells	Length of Shells (in.)
		Total Length	

Answer _____

Analyze Victor solved this problem by finding the length of 4 scallops, 2 sand dollars, and 1 conch. Then he multiplied the answer by 10. Does this method work?

_____

_____

40 Unit 1 **Using Whole Numbers and Decimals**

Practice

Solve the problems. Show your work.

6 Neil and Ned have different beach towels. Neil's towel is 150.9 centimeters by 92.2 centimeters. Ned's towel is 158.3 centimeters by 87.9 centimeters. Each boy thinks his towel is bigger. Whose towel has the greater area? How much greater?

Answer _____

Infer Is it possible for two towels with different lengths and widths to have the same area? Explain.

_____

_____

7 Ms. Sherrill runs a food cart at the beach. She makes a $0.13 profit on each frozen yogurt she sells. Her profit for a salad is $0.48. Ms. Sherrill also makes $0.37 profit on each turkey wrap she sells. Today, she sold 18 frozen yogurts, 23 salads, and 29 turkey wraps. What is Ms. Sherrill's total profit?

Answer _____

Discuss How would you label a table to solve the problem?

_____

_____

Create Look back at Problem 5. Change the numbers of each kind of shell. Write and solve a problem about the garden. Be sure your problem can be solved by using the strategy *Make a Table*.

Lesson 3 **Strategy Focus: Make a Table** 41

Lesson 4

Strategy Focus
Guess, Check, and Revise

MATH FOCUS: Divide Decimals

Learn

Read the Problem

Jason's favorite food in the science museum's café is the astronaut meatloaf. One pouch of astronaut meatloaf costs $1.19. Jason has $20 to spend. He wants to buy as many pouches of meatloaf as he can. How many pouches of meatloaf can Jason buy with $20?

Reread Look for details that will help you answer the question.

- What is this problem about?
  _____

- What kinds of information do I know?
  _____

- What do I need to find?
  _____

Mark the Text →

Search for Information

As you read the problem again, mark important facts. Underline the question you need to answer.

Record Write the details from the problem.

Jason wants to buy _____ .

Each pouch of meatloaf costs _____ .

Jason has _____ to spend.

Think about how this information can help you choose a problem-solving strategy.

42   Unit 1   Using Whole Numbers and Decimals

Decide What to Do

You know what Jason wants to buy and how much money he has.

Ask How can I find how many pouches of meatloaf Jason can buy with $20?

- I can use the strategy *Guess, Check, and Revise*.
- I can start by guessing a quotient. Next, I can multiply my guess by the cost of one pouch of meatloaf. Then I can compare that product to $20. I can revise my guesses until I find the greatest number of pouches Jason can buy with $20.

Use Your Ideas

Step 1 Make a guess of how many pouches Jason can buy.

The cost of a pouch is $0.19 more than $1, so a guess of 20 is _____ .

Start with a lesser number, such as 17. Write the number in the first column of the table. Multiply your guess by $1.19, the cost of a pouch of meatloaf. Then compare the product to $20.

Guess	Guess × $1.19	Compare to $20
17	$20.23	$20.23 is too high.
15	$17.85	

You can use a table to keep track of your guesses for the quotient.

Step 2 Your first guess is too high. Try 15. Fill in the row in the table for 15. The difference between $20 and $17.85 is $2.15. That product is too low.

Step 3 Your second guess is _____ . Try 16.

Jason can buy _____ pouches of meatloaf with $20.

Review Your Work

Check that Jason cannot buy another pouch of meatloaf.

Explain How do you know that your answer is correct?

_____

_____

Try

Solve the problem.

(1) Devin learned that the average American eats about 140 pounds of potatoes each year. About 94 potato plants produce enough potatoes each year for the average American. About how many pounds of potatoes does a potato plant produce? Round your answer to the nearest tenth of a pound.

Mark the Text

Read the Problem and Search for Information

What operation will you use to solve this problem?

Decide What to Do and Use Your Ideas

The strategy *Guess, Check, and Revise* can help you even if you do not know how to divide with decimals.

Step 1 Make a guess for the number of pounds of potatoes. Try 1. Write your guess in the table. Multiply by 94, the number of plants. Compare the product to 140 pounds.

Guess	Guess × 94	Compare to 140 pounds
1 pound	94 pounds	94 pounds is too low.
1.3 pounds		

Ask Yourself

What is a good next guess for the quotient?

Step 2 Your guess of 1 is _____. So try a greater number. Try 1.3. Fill in the row in the table for _____ .

Step 3 Your guess of 1.3 is still too low. Try 1.6.

Step 4 Your guess of 1.6 is _____. Try 1.5. _____ is close to 140 pounds.

So a potato plant produces about _____ pounds of potatoes.

Review Your Work

Check that you multiplied correctly to find the products.

Describe How did you use one guess to help make your next guess?

_____

_____

Unit 1 **Using Whole Numbers and Decimals**

Apply

Solve the problems.

(2) Emma fills a new aquarium tank that holds 1,350 gallons of water. Her hose pours 18.4 gallons per minute into the tank. About how many minutes will it take Emma to fill the tank? Round your answer to the nearest minute.

Guess	Guess × 18.4	Compare to 1,350 gallons
100 minutes	1,840 gallons	

> **Ask Yourself**
> If I round the numbers in the problem to 1,500 gallons of water and 15 gallons per minute, what do I get for a first guess?

◀ **Hint** Compare your product to 1,350 gallons.

Answer _____

Identify How does the table help you keep track of your guesses?

_____

_____

(3) Kimi's favorite sports car car gets 14 miles per gallon. She imagines she is riding in it from her house to her grandparents' house 102 miles away. About how many gallons of gas would it take to make the trip? Round your answer to the nearest tenth of a gallon.

Guess	Guess × 14	Compare to 102 miles
7 gallons		

> **Ask Yourself**
> What should be my first guess?

◀ **Hint** Use the result from your first guess to help make your second guess.

Answer _____

Recognize Why do you need to compare the product to 102 miles every time you make a guess?

_____

_____

Lesson 4 **Strategy Focus: Guess, Check, and Revise**   45

Ask Yourself
Is the quotient that I guessed too high or too low?

Hint When the divisor and the dividend are close to being compatible numbers, you can make a good guess at the quotient.

④ Troy learned that a planet's day is the length of time it takes for the planet to spin once on its axis. One day on Venus lasts 2,802 Earth hours. Jupiter spins much more quickly. One day on Venus is the same as about 283 days on Jupiter. About how many Earth hours long is one day on Jupiter? Round your answer to the nearest tenth.

Guess	Guess × _____	Compare to _____
10 hours		

Answer _____

Compare How is guessing quotients and checking each guess by multiplying and comparing different when decimals are used?

_____

_____

Hint First find out how many bowls Ronde can fill.

⑤ At the science museum, Ronde mixes cornstarch and water to make goo. He shows how the goo can act like a solid or a liquid in different situations. Ronde has 20 liters of goo. He fills bowls with 0.35 liter of goo. The last bowl is not completely filled. How much extra goo does Ronde need to fill the last bowl?

Ask Yourself
I need to find out how much extra goo Ronde needs for the last bowl. How much goo is in the last bowl?

Answer _____

Analyze Dee's answer is 0.05 liter. What mistake might she have made?

_____

_____

46 Unit 1 **Using Whole Numbers and Decimals**

Practice

Solve the problems. Show your work.

6 Jess went to a jelly bean museum. She saw a giant jar of jelly beans that holds 100 pounds, or 1,600 ounces, of jelly beans. There were 32 different flavors of jelly beans. Jess wondered how many days it would take to empty the jar if she ate 3.5 ounces of jelly beans every day. For how many days would Jess eat jelly beans before they were all gone?

Answer _____

Determine What information is not needed to solve the problem?

_____

_____

7 Mr. Marin teaches at an art museum. He teaches his students how to make a bracelet with copper wire. He needs to cut 7.5-inch pieces of wire. He has 130 inches of wire on a spool. He cuts as many 7.5-inch pieces as he can. About how much wire will be left on the spool? Round your answer to the nearest tenth.

Answer _____

Examine How is this problem similar to Problem 5?

_____

_____

Create

Look back at Problem 3. Write and solve a problem using a new value for gas mileage and the miles for the trip. Be sure your problem can be solved using the strategy *Guess, Check, and Revise*.

Lesson 4 **Strategy Focus: Guess, Check, and Revise** 47

UNIT 1 Review

In this unit, you worked with four problem-solving strategies. You can often use more than one strategy to solve a problem. So if a strategy does not seem to be working, try a different one.

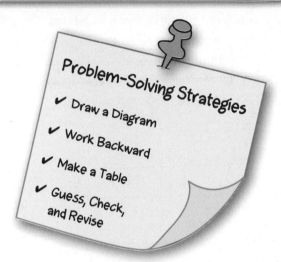

Problem-Solving Strategies
- ✔ Draw a Diagram
- ✔ Work Backward
- ✔ Make a Table
- ✔ Guess, Check, and Revise

Solve each problem. Show your work. Record the strategy you use.

1. Nigel and Rob collect rocks. Rob's rocks have a total mass of 53.2 grams more than Nigel's. Altogether, their rocks have a mass of 342.5 grams. How many grams of rocks does Rob have?

   Answer _____

   Strategy _____

2. Jake saves coins in a jar. He takes out $8.25 to go to the movies. He comes home and puts $0.39 back in the jar. He counts the money and has $27.35. How much did Jake have in the jar before taking out money for the movies?

   Answer _____

   Strategy _____

3. Ariel makes and sells jewelry. She earns $5.35 for each necklace she sells, $2.82 for each bracelet, and $3.89 for each pair of earrings. Ariel has sold 3 necklaces, 5 bracelets, and 4 pairs of earrings. How much money has she earned?

Answer _____

Strategy _____

4. Amy, Matt, and Rory all grow tomatoes. By the end of the season, Matt had grown 18.3 fewer pounds of tomatoes than Amy. Rory grew 9.6 fewer pounds than Matt. Rory grew 50 pounds of tomatoes. How many pounds of tomatoes did Amy grow?

Answer _____

Strategy _____

5. Carol had $20. She met Kelly at the mall and bought a smoothie. Kelly paid Carol back the $3.25 she had borrowed from Carol last week. Then Carol bought a belt for $7.95. If Carol had $13.41 left, how much did the smoothie cost?

Answer _____

Strategy _____

Explain what information in the problem made you choose a strategy for solving the problem.

49

Solve each problem. Show your work. Record the strategy you use.

6. There are 123 students at a local school. 42 students play basketball. 71 students play soccer. 12 students play both soccer and basketball. How many students do not play soccer or basketball?

   Answer _____

   Strategy _____

7. Dante made a tower that was 141 centimeters tall out of empty containers. He used cans that are 10.2 centimeters tall, butter tubs that are 8.8 centimeters tall, and jars that are 16.6 centimeters tall. He used 3 cans and 4 jars. How many butter tubs did Dante use?

   Answer _____

   Strategy _____

8. Jon and Alex collected pond water for a science project. Jon collected three times as much water as Alex. Together, they collected 2.6 liters. How much pond water did Jon collect?

   Answer _____

   Strategy _____

   Explain how your strategy helped you choose the operations you needed to solve the problem.

9. Becky has $1.29 in a jar. She has pennies, nickels, dimes, and quarters. Becky has the same number of quarters and dimes and the same number of nickels and pennies. How many of each coin are in the jar?

10. Dean spent $6 buying marbles. He bought twice as many small marbles as large marbles. The small marbles cost $0.20 each and the large marbles cost $0.35 each. How many of each type of marble did Dean buy?

Answer _____

Strategy _____

Answer _____

Strategy _____

Write About It

Look back at Problem 9. Describe how you found your answer.

_____

_____

_____

Team Project: Stock a Store

Your school is opening a store that will sell school supplies. Your team is in charge of ordering the items for the store's grand opening. The supplies are shown in the table. You have a budget of $50.

Plan
1. You must choose at least one of each item.
2. You cannot spend more than your budget.

Decide As a group, choose what items you will order. Think about what students at your school will need most.

Organize Keep track of what you decided to order and be sure that you did not spend more than your budget.

Present As a team, present your plan to the class. Tell what you ordered and the costs. Explain how you made your decisions.

Item	Price
4 dozen pencils	$3.49
1 dozen pens	$2.49
1 dozen erasers	$1.79
1 dozen notepads	$5.99
1 pack of 10 folders	$3.79

UNIT 2: Problem Solving Using Fractions

Unit Theme: Structures

Structures can be huge. They can also be small. Bridges and famous monuments are structures. In this unit, you will see how math is used to build and measure the structures in the world around you.

Math to Know

In this unit, you will apply these math skills:

- Fraction and mixed number concepts
- Add and subtract fractions and mixed numbers
- Multiply and divide fractions and mixed numbers

Problem-Solving Strategies

- Use Logical Reasoning
- Look for a Pattern
- Write an Equation
- Solve a Simpler Problem

Link to the Theme

Write another paragraph about Pia's presentation. Include some of the facts from the table at the right.

Pia is giving an oral presentation about bridges. She tells the class some details about famous bridges around the world.

_____

_____

_____

_____

_____

_____

Bridge Name	Country	Length (miles)
Golden Gate	United States	$1\frac{7}{10}$
Penang	Malaysia	$8\frac{2}{5}$
Vasco da Gama	Portugal	$10\frac{7}{10}$
Donghai	China	$20\frac{1}{5}$

Use Math Language

Review Vocabulary

The list below shows vocabulary terms in this unit. Knowing the meaning of these terms will help you understand the problems.

denominator	equation	numerator	prime number
dividend	mixed number	pattern	reciprocal
divisor	multiple	perimeter	variable

Vocabulary Activity   Prefixes

Look for a prefix at the beginning of a math term. It can give you a clue to the word's meaning. Use words from the list to complete the following sentences.

1. The prefix *equ-* means same or equal. _____ begins with the prefix *equ-*.

2. The prefix *multi-* means many. _____ has the same prefix as the word *multiply*.

3. The prefix *peri-* means around. The distance around a plane figure is called the _____ .

4. The prefix *vari-* means different. _____ begins with the prefix *vari-*.

Graphic Organizer   Word Web

Complete the graphic organizer.

- In the top oval, write a definition of *reciprocal*.

- In each linked oval, write a different vocabulary term related to reciprocal and then write a definition of each related term.

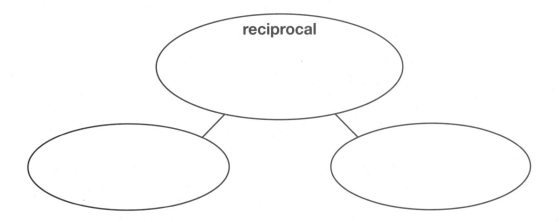

Lesson 5

Strategy Focus
Use Logical Reasoning

MATH FOCUS: Number and Fraction Concepts

Learn

Read the Problem

Mr. Ruiz is making a model rocking chair, easy chair, recliner, and kitchen chair. The denominator of the rocking chair height is half the denominator of the recliner height. The denominator of the rocking chair height is equal to the denominator of the easy chair height. The easy chair is shortest. Match each type of chair with its height.

Chair Heights
$1\frac{7}{8}$ in.
$1\frac{5}{8}$ in.
$2\frac{7}{16}$ in.
$2\frac{1}{4}$ in.

Reread Ask yourself questions while you read.

- What is the problem about?
  _____

- What information is given?
  _____

- What am I asked to do?
  _____

Mark the Text

Search for Information

Look for data and words to use in solving the problem.

Record Write the data and words that you identified.

The chair heights are _____ , _____ , _____ ,

and _____ inches.

The denominator of the rocking chair height is _____ the denominator of the recliner height.

The denominator of the _____ height is equal to the denominator of the easy chair height.

The _____ is the shortest.

Use the clues and data to choose a problem-solving strategy.

54 Unit 2 Using Fractions

Decide What to Do

You know the heights of four different kinds of model chairs.

Ask How can I match each type of chair with its height?

- I can use the strategy *Use Logical Reasoning*. I can use the clues and make a table to eliminate choices.

Use Your Ideas

Step 1   You know that the **denominator** of the rocking chair height is half the denominator of the recliner height. So the rocking chair height does not have 16 as its denominator. Use an X to rule out $2\frac{7}{16}$ inches. The recliner cannot be $2\frac{1}{4}$ inches tall. Mark this space with an X.

	$1\frac{7}{8}$ in.	$1\frac{5}{8}$ in.	$2\frac{7}{16}$ in.	$2\frac{1}{4}$ in.
Rocking Chair			X	X
Easy Chair			X	X
Recliner				X
Kitchen Chair				✔

Use an X to show you have eliminated a choice and a ✔ to show the right choice.

Step 2   You know that the denominator of the rocking chair height is equal to the denominator of the easy chair height. So neither the rocking chair nor the easy chair can be $2\frac{7}{16}$ inches or $2\frac{1}{4}$ inches tall. The only chair that can be $2\frac{1}{4}$ inches tall is the kitchen chair. Mark this space with a ✔.

Step 3   You know that the easy chair is $1\frac{5}{8}$ inches tall. Mark this space with a ✔. The height of the rocking chair is $1\frac{7}{8}$ inches. Mark this with a ✔. Mark the height of the recliner with a ✔.

So the height of the rocking chair is _____ , the height of

the easy chair is _____ , the height of the recliner is

_____ , and the height of the kitchen chair is _____ .

Review Your Work

Check that your answers fit all of the clues.

Describe How did the table help you find your answer?

_____

_____

Try

Solve the problem.

(1) Vivien is using a precise machine to cut a metal strip. The strip is between 7 and 8 inches wide. The width is a mixed number in simplest form. The denominator of the fraction is a multiple of 20 and less than 100. The numerator is 27 less than the denominator. What are the two possible widths of the strip?

Mark the Text

Read the Problem and Search for Information

Reread the problem and circle important words and numbers.

Decide What to Do and Use Your Ideas

You can use the strategy *Use Logical Reasoning* for this problem.

Step 1   List the multiples of 20 that are less than 100.

20, 40, _____ , _____

Step 2   Find possible numerators.

The numerator is _____ less than the denominator.
So the denominator cannot be 0 or 20.

40 − 27 = _____

60 − 27 = _____

80 − 27 = _____

Ask Yourself

What does it mean for a fraction to be in simplest form?

Step 3   Write numbers between 7 and 8 using the fractions.

$7\frac{}{40}, 7\frac{}{60}, 7\frac{}{80}$

Which fraction is not written in simplest form? _____

So the possible widths of the strip are _____ and _____ inches.

Review Your Work

Make sure that your answer makes sense.

Recognize   Tara says that another possible answer is $7\frac{73}{100}$ inches. Why is this an incorrect answer?

_____

_____

Unit 2   Using Fractions

Apply

Solve the problems.

2 Hiroshi has three sizes of screws. The screws are in boxes labeled A, B, and C. The diameters of the screws are $\frac{1}{8}$ inch, $\frac{5}{32}$ inch, and $\frac{7}{64}$ inch. The screws in box C are smaller than the screws in box B. The screws in box A are the largest. What size screw is in each box?

	$\frac{7}{64}$ inch	$\frac{1}{8}$ inch	$\frac{5}{32}$ inch
A			
B			
C			

◀ **Hint** Arrange the diameters from least to greatest in the table.

Ask Yourself

Does it matter which clue I use first?

Answer _____

Explain Why is it helpful to arrange the diameters from least to greatest in the table?

_____

_____

3 Jan is building a wheelchair ramp. It is 3 feet wide. The ramp is less than 15 feet high. The ratio of the height of the ramp to its length is 1 to 12. The height of the ramp is a prime number of feet. The length is a multiple of 10 feet. What are the height and length of the ramp?

Ask Yourself

What are possible lengths for the ramp?

Prime numbers less than 15: 2, 3, _____, _____, _____, _____

Multiples of 10: 10, 20, _____, _____, _____, _____

◀ **Hint** Find a ratio equivalent to 1:12.

Answer _____

Identify What information is given that is *not* needed to solve the problem?

_____

_____

Lesson 5 **Strategy Focus: Use Logical Reasoning** 57

Ask Yourself

Which length can you immediately eliminate for the bathroom from the first clue?

Hint Organize the information in a table so you can eliminate choices.

(4) Patty decorates the floors of her entryway, kitchen, bathroom, and hallway with tiles. She uses $2\frac{1}{2}$-inch, $4\frac{2}{3}$-inch, $6\frac{3}{8}$-inch, and $10\frac{5}{6}$-inch square tiles. The bathroom tiles are not the smallest. The kitchen tiles are larger than the bathroom tiles. The denominator of the tiles in the bathroom is the same as the numerator of the tiles in the hallway. Which size tiles does Patty use for the entryway?

Entryway				
Kitchen				
Bathroom				
Hallway				

Answer _____

Determine How did you know the kitchen tiles were not $2\frac{1}{2}$ inches?

_____

_____

Hint Write 0.5 as a fraction.

Ask Yourself

What does it mean for a number to be prime?

(5) Jonah has a curtain rod that is less than one inch in diameter, but greater than 0.5 inch. When the diameter is written as a fraction in simplest form, the numerator is a prime number. The denominator is a multiple of 2 and less than 10. The numerator is 3 less than the denominator. What is the diameter of the curtain rod?

Multiples of 2: _____ , _____ , _____ , _____

Prime numbers: _____ , _____ , _____ , _____

Answer _____

Extend What is another clue that could be used in this problem?

_____

_____

58 Unit 2 **Using Fractions**

Practice

Solve the problems. Show your work.

6 Kevin bought 4 throw rugs. The rugs are brown, white, green, and red. The lengths of the rugs are $4\frac{3}{4}$ feet, $5\frac{1}{6}$ feet, $3\frac{1}{4}$ feet, and $4\frac{5}{12}$ feet. The length of the brown rug has the least fractional part. The white rug is shorter than the green rug, but is not the shortest. What are the lengths of the rugs?

Answer _____

Develop What is another question you could ask about the problem?

_____

_____

7 The gauge of a wire is the measure of its diameter. Jackie needs 13-gauge wire. The wire is between 1.7 and 1.85 millimeters in diameter. When written as a mixed number in simplest form, the denominator of the fraction is prime. The numerator is 1 less than the denominator. What is the diameter of 13-gauge wire?

Answer _____

Conclude What are two other mixed numbers you tried? Why did you eliminate them?

_____

_____

Create Look back at Problem 4. Change at least one length and one of the clues. Write a new problem that can be solved using the strategy *Use Logical Reasoning*. Solve your problem.

Lesson 6

Strategy Focus
Look for a Pattern

MATH FOCUS: Add and Subtract Fractions

Learn

Read the Problem

Jeff is walking along the bridge shown. The vertical lines on the diagram are posts. The posts are equally spaced and the distance from the start and the end of the bridge to the nearest post is the same. Jeff keeps track of the distance he walks. He starts on the left side of the bridge and walks $3\frac{3}{4}$ feet to the first post. When he gets to the second post, he has walked $7\frac{1}{2}$ feet. When he gets to the third post, he has walked $11\frac{1}{4}$ feet. How far does Jeff walk from start to end?

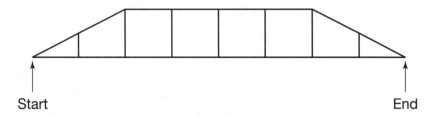

Reread Think of these questions as you read the problem.

- What is the problem about?
  _____

- What does the diagram show?
  _____

- What do I need to find?
  _____

Mark the Text

Search for Information

Study the problem.

Record List the information you need. Mark the distances on the diagram.

Distance to first post: _____

Distance to second post: _____

Distance to third post: _____

Use this information to help you look for a pattern.

60 Unit 2 **Using Fractions**

Decide What to Do

You know that Jeff walks all the way across the bridge. You also know distances from the start to some posts.

Ask How can I find how far Jeff walks from start to end?

- I can find the distances between the posts and use the strategy *Look for a Pattern*.
- I can decide how far I should extend the pattern. Then I can find the distance from start to end.

Use Your Ideas

Step 1 Find the distance from the first post to the second post and from the second post to the third post.

$7\frac{1}{2} - 3\frac{3}{4} =$ _____ $11\frac{1}{4} - 7\frac{1}{2} =$ _____

Those distances are the same as from the beginning to the first post.

So the pattern is *add* _____ .

Step 2 Extend the pattern.

You know the distance to the third post. There are a total of 7 posts. Then Jeff walks to the end. You need to extend the pattern.

Post	Start	1	2	3	4	5	6	7	End
Distance (feet)	0	$3\frac{3}{4}$	$7\frac{1}{2}$	$11\frac{1}{4}$	15	$18\frac{3}{4}$			

$+3\frac{3}{4} \quad +3\frac{3}{4} \quad +3\frac{3}{4} \quad +3\frac{3}{4} \quad +3\frac{3}{4} \quad +3\frac{3}{4} \quad +3\frac{3}{4} \quad +3\frac{3}{4} \quad +3\frac{3}{4}$

Remember, Jeff walks to the end of the bridge, not to the last post.

So Jeff walks _____ feet from start to end.

Review Your Work

Make sure you have added correctly as you completed the table.

Recognize Kia says that you can find the answer by multiplying the distance to the fourth post by 2. Is this correct? Explain.

_____

_____

Try

Solve the problem.

(1) Wildlife tunnels can help reduce the number of accidents between cars and wild animals. One town makes a tunnel using metal tubes. Each tube is $5\frac{1}{3}$ feet long. There is a 6-inch connecting piece between each tube. How many tubes and connecting pieces are needed to make a tunnel $22\frac{5}{6}$ feet long?

Mark the Text

▢ Read the Problem and Search for Information

Reread the problem. Circle important words and numbers.

▢ Decide What to Do and Use Your Ideas

You can use the strategy *Look for a Pattern* to find the answer.

Write the inches given in the problem as fractional parts of feet.

Step 1 Identify the pattern you will use.

Each tube is _____ feet long.

Each connecting piece is _____ foot long.

So the pattern is *add* _____ , *add* _____ .

Ask Yourself

Will there be connecting pieces at the ends of the tunnel?

Step 2 Use the pattern to make a table showing the tunnel lengths using different numbers of tubes.

Piece	Tube 1	Connecting Piece	Tube 2	Connecting Piece	Tube 3	Connecting Piece	Tube 4
Length (feet)	$5\frac{1}{3}$	$5\frac{5}{6}$	$11\frac{1}{6}$	$11\frac{2}{3}$			

$+\frac{1}{2}$ $+5\frac{1}{3}$ $+\frac{1}{2}$ _____ _____ _____

So it will take _____ tubes and _____ connecting pieces to make a tunnel $22\frac{5}{6}$ feet long.

▢ Review Your Work

Make sure you used the right number of tubes and pieces.

Summarize How did a table help you solve the problem?

_____

_____

62 Unit 2 **Using Fractions**

Apply

Solve the problems.

2 Melanie uses planks to build a footbridge. She alternates between planks that are $6\frac{1}{4}$ inches and $7\frac{3}{8}$ inches wide. How long is the footbridge after Melanie has laid 6 planks?

Ask Yourself
What width is the third plank?

Hint You may want to write the width of each plank on the diagram.

Plank	1	2	3	4	5	6
Total Length (in.)	$6\frac{1}{4}$	$13\frac{5}{8}$				

Answer _____

Explain How did you use a pattern to find your answer?

_____

3 Giant tunnel-boring machines (TBMs) were used to build a water-supply tunnel in Greece. The tunnel is about 18 miles long. An engineer measured how far the TBM had drilled over several hours. The TBM drilled $4\frac{11}{12}$ feet in 1 hour, $9\frac{5}{6}$ feet in 2 hours, $14\frac{3}{4}$ feet in 3 hours, and $19\frac{2}{3}$ feet in 4 hours. At that rate, how far did the machine drill in 7 hours?

Hint Find the number of feet the TBM drills each hour.

Ask Yourself
How should I set up my table to show the pattern?

Hours	1	2	3	4	5	6	7
Distance (ft)	$4\frac{11}{12}$						

Answer _____

Examine What information is given that is *not* needed?

_____

Lesson 6 **Strategy Focus: Look for a Pattern** 63

4 Mr. Young is putting safety lights in a 59-foot work tunnel. He puts the first light at the start of the tunnel. He puts the next one $12\frac{1}{3}$ feet away. He puts the third light $11\frac{1}{3}$ feet away from the second. He places the fourth light $10\frac{1}{3}$ feet away from the third. If this pattern continues, how many lights will Mr. Young use to cover 59 feet?

Ask Yourself: What is the pattern in the distances?

Number of Lights	2	3	4			
Distance from Previous Light (feet)	$12\frac{1}{3}$	$11\frac{1}{3}$				
Total Distance (feet)	$12\frac{1}{3}$	$23\frac{2}{3}$				

Hint Do not forget the first light at the beginning of the tunnel.

Answer _____

Analyze Why does the table begin with two lights?

_____

_____

5 A TBM drills $120\frac{3}{4}$ feet in 24 hours. At that rate, will the TBM be able to drill 2,500 feet in 3 weeks? If so, how much farther will it drill? If not, how many more feet will it have to drill?

Ask Yourself: If I find how far the TBM can drill in 7 days, how can I use that to find how far it can drill in 3 weeks?

Hint Make a second table for 3 weeks using the distance the TBM can drill in 1 week.

Answer _____

Relate What is another question you could ask about the problem?

_____

_____

64   Unit 2   Using Fractions

Practice

Solve the problems. Show your work.

6 Mrs. Von is checking the emergency phones in a tunnel. The tunnel is $2\frac{1}{2}$ miles long. The first phone is $\frac{1}{4}$ mile from the start of the tunnel and the last phone is $\frac{1}{4}$ mile from the end of the tunnel. The second is $\frac{1}{2}$ mile from the start of the tunnel. The third is $\frac{3}{4}$ mile from the start of the tunnel. The fourth phone is 1 mile from the start of the tunnel. If this pattern continues, how many phones will Mrs. Von check?

Answer _____

Determine Abby says that Mrs. Von will check 11 phones. What mistake might Abby have made?

_____

_____

7 A carpenter cuts strips of wood from a 2-foot board. The wood that the blade touches turns into sawdust, removing a section about $\frac{1}{4}$ inch wide. What is the greatest number of 1-inch strips the carpenter can cut from the wood?

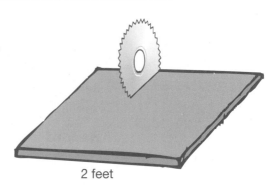

2 feet

Answer _____

Justify How did you know when you had reached the greatest number of strips that could be cut?

_____

_____

Create

Look back at Problem 1. Change the length of a tube and the length of the tunnel. Write a problem that can be solved using the strategy *Look for a Pattern*. Solve your problem.

Lesson 7

Strategy Focus
Write an Equation

MATH FOCUS: Multiply Fractions

Learn

Read the Problem

The tallest Ferris wheel in the world used to be in Japan. In 2008, a taller Ferris wheel was built in Singapore. It is called the Singapore Flyer and is about 541 feet tall. The Ferris wheel in Japan is $\frac{7}{10}$ the height of the Singapore Flyer. To the nearest foot, how tall is the Ferris wheel in Japan?

Reread Ask yourself questions as you read the problem.

- What is the problem about?

  _____

- What does the problem ask?

  _____

Search for Information

Mark the Text →

Reread the problem. Think about the numbers and words you will need to answer the question.

Record Write the facts you need.

The Singapore Flyer is _____ feet tall.

The height of the Ferris wheel in Japan is _____ the height of the Singapore Flyer.

You can use these facts to help you decide how to solve the problem.

66 Unit 2 **Using Fractions**

Decide What to Do

You know the height of one Ferris wheel. You know how the height of the other Ferris wheel relates to that first height.

Ask How can I find the height of the Ferris wheel in Japan?

- I can write what I know about the relationship between the heights using words.

- I can use the strategy *Write an Equation*. I can write an equation using the values I know to find the value I do not know.

Use Your Ideas

Step 1 Write an equation using words to show how the heights are related.

height of wheel in Japan = $\frac{7}{10}$ × height of Singapore Flyer

Step 2 Write an equation. Choose a variable to represent the height you do not know. Substitute the height you know.

Let h stand for the height of the Ferris wheel in Japan.

$h = \frac{7}{10} \times$ _____

Step 3 Solve your equation. Multiply the numerators. Multiply the denominators. Write the product in simplest form.

$h = \frac{7}{10} \times$ _____

$h =$ _____

$h =$ _____

> Write the whole number as a fraction.

To the nearest foot, the Ferris wheel in Japan is _____ feet tall.

Review Your Work

Check your work. Does it make sense that the height you found is less than the height you are given?

Identify How did you know which operation you would use?

_____

_____

Try

Solve the problem.

1) A new building in a city in Canada is about $453\frac{1}{2}$ feet tall. Nearby is the CN Tower. The CN Tower is about 4 times as tall as the new building. About how tall is the CN Tower?

■ Read the Problem and Search for Information

Identify the information in the problem that you will need to use. Think about how the heights relate to each other.

■ Decide What to Do and Use Your Ideas

Write and solve an equation that shows the relationship between the heights.

Ask Yourself

What do I need to find in this problem? What variable can I use to stand for that value?

Step 1 Write a word equation.

height of CN Tower = 4 × height of new building

Step 2 Write an equation. Use a variable to represent what you do not know. Substitute the height you know.

Let h stand for the height of the CN Tower.

$h = 4 \times$ _____

Step 3 Solve the equation.

$h =$ _____

So the height of the CN Tower is about _____ .

■ Review Your Work

Check your computation to be sure you used the correct numbers in the equation.

Examine How did writing the equation in words help you solve the problem?

_____

_____

Unit 2 Using Fractions

Apply

Solve the problems.

2 Owen learns that the Dublin Spire in Ireland is about 7 times as tall as the post office on the same street. The post office is about $56\frac{1}{4}$ feet tall. About how tall is the Dublin Spire?

height of Dublin Spire = _____ × height of post office

Let h stand for the height of the Dublin Spire.

$h =$ _____ ◯ _____

> **Ask Yourself**
> How are the heights of the buildings related?

> ◀ **Hint** An estimate can help you check that your answer makes sense.

Answer _____

Differentiate Would your equation change if the problem stated that the post office is about $\frac{1}{7}$ the height of the Dublin Spire? Why or why not?

_____

_____

3 The tallest bridge in the world is in France. The tallest bridge in the United States is the Golden Gate Bridge. The Golden Gate Bridge is about 746 feet tall. The height of the bridge in France is about $1\frac{1}{2}$ times the height of the Golden Gate Bridge. About how tall is the bridge in France?

Let h stand for the height of the bridge in France.

$h =$ _____ ◯ _____

> **Ask Yourself**
> Would writing a word equation first be helpful?

> ◀ **Hint** Rewrite the whole number in the equation as a fraction.

Answer _____

Relate How did you know which operation to use?

_____

_____

Lesson 7 **Strategy Focus: Write an Equation**

④ The longest bridge in the United States is in Louisiana. It is about 126,000 feet long. A bridge in Virginia is about $\frac{5}{8}$ as long as the bridge in Louisiana. A bridge in Michigan is about $\frac{1}{3}$ as long as the bridge in Virginia. About how long is the bridge in Michigan?

Ask Yourself
How can use what I know about the bridge in Louisiana to find the length of the bridge in Virginia?

Hint You can solve the problem by multiplying by fractions.

Let l stand for the length of the bridge in Louisiana.

$l =$ _____ feet

Let v stand for the length of the bridge in Virginia.

$v = \frac{5}{8} \times l$

$v =$ _____ feet

Let m stand for the length of the bridge in Michigan.

$m = \frac{1}{3} \times v$

Answer _____

Analyze Can you write one equation to solve this problem? Explain why or why not.

_____

_____

⑤ The tallest building in Mississippi is about $346\frac{1}{4}$ feet tall. The tallest building in the United States is about 4 times that height. The tallest building in the world is about $1\frac{7}{8}$ times the height of the tallest building in the United States. To the nearest hundred feet, how tall is the tallest building in the world?

Hint The final answer is the tallest of the three buildings in the problem.

Let m stand for the height of the tallest building in Mississippi.

$m =$ _____ feet

Ask Yourself
What equations will help me to answer this question?

Let u stand for the height of the tallest building in the United States.

$u =$ _____ $\times m$

Let w stand for the height of the tallest building in the world.

$w =$ _____

Answer _____

Conclude Can you find the exact height of the tallest building in the world from this problem? Explain why or why not.

_____

_____

70   Unit 2   **Using Fractions**

Practice

Solve the problems. Show your work.

6 The original Ferris wheel was built for the World's Fair in Chicago. In 2010, the Beijing Great Wheel is expected to open. It will be 684 feet tall. The original Ferris wheel was about $\frac{2}{5}$ of that height. How tall was the original Ferris wheel to the nearest foot?

Answer _____

Consider How would writing an equation help you represent the problem?

_____

7 Bridge A is about $1,203\frac{1}{4}$ feet from one support to the next. The distance between supports on Bridge B is about 3 times as far. Bridge C has a distance between supports about $\frac{3}{8}$ as long as the distance between supports on Bridge B. To the nearest 10 feet, what is the distance between supports on Bridge C?

Answer _____

Explain Is this problem similar to another problem in this lesson? Explain how it is similar and how it is different.

_____

Create Look back at Problem 1. Describe the heights of 2 buildings. Write a problem that relates the heights as a fraction. Be sure your problem can be solved using the strategy *Write an Equation*. Solve your problem.

Lesson 7 **Strategy Focus: Write an Equation** 71

Lesson 8

Strategy Focus
Solve a Simpler Problem

MATH FOCUS: Divide Fractions

Learn

Read the Problem

Sally is visiting the Gateway Arch in Saint Louis. It is 630 feet tall. The distance between the two bases of the arch is also 630 feet. Sally walks from one base of the arch to the other base in a straight line. Each step Sally takes is $2\frac{1}{2}$ feet long. How many steps does Sally take to get from one base of the arch to the other?

Reread Ask yourself questions as you read the problem.

- What is the problem about?
  _____

- What measurements are given in the problem?
  _____

- What do I need to find?
  _____

Mark the Text

Search for Information

Read the problem again.

Record Write the data you will need to solve this problem.

The bases of the arch are _____ feet apart.

Each of Sally's steps is _____ feet long.

You can use what you know to decide how to solve the problem.

72 Unit 2 **Using Fractions**

Decide What to Do

You know the distance from one base of the Gateway Arch to the other. You know how far Sally walks with each step she takes.

Ask How can I find the number of steps that Sally takes to get from one base to the other?

- I need to divide the total distance by the length of Sally's step. But the length of her step is a mixed number.
- I can use the strategy *Solve a Simpler Problem*. Instead of starting with these numbers, I will use easier numbers first to make sure I am solving correctly.

Use Your Ideas

Step 1 Find simpler numbers to solve the problem.

630 feet is about _____ feet.

$2\frac{1}{2}$ feet is about _____ feet.

Step 2 Solve the problem with the simpler numbers you chose.

I will divide, so _____ ÷ _____ = _____ feet.

Step 3 Change the mixed number divisor to a fraction. Then multiply the dividend by the reciprocal of the divisor.

$630 \div 2\frac{1}{2} = ?$

$630 \div \text{———} = ?$

$630 \times \frac{2}{5} = $ _____

Now use the numbers given in the problem and solve.

Sally takes _____ steps to get from one base of the arch to the other.

Review Your Work

Check that the number of steps is reasonable.

Describe Why is solving a simpler problem a good strategy to use to solve this problem?

_____

_____

73

Try

Solve the problem.

1. The Washington Monument in Washington, D.C. has a square base. The perimeter of the base is about $220\frac{1}{2}$ feet. About what length is one side of the base of the Washington Monument?

Mark the Text

■ Read the Problem and Search for Information

Circle the information in the problem. Identify what you need to find.

■ Decide What to Do and Use Your Ideas

I can use the strategy *Solve a Simpler Problem*.

I know the _____ of the base of the monument, but I need to find the length of _____ .

Ask Yourself

How would I solve the problem if the perimeter was 12 feet?

Step 1 Solve a simpler problem.

A square has _____ equal sides.

If the perimeter of a square is 12 feet, then the length of 1 side is _____ .

The equation that shows this is 12 ◯ _____ = _____ .

Step 2 Use the same method to solve the real problem.

You need to divide the perimeter by _____ .

Write the equation. Then solve it.

_____

The length of one side of the base of the Washington Monument is

about _____ .

■ Review Your Work

You can multiply your answer by 4 to check your work.

Explain Why did you use a division equation in Step 2?

_____

_____

74 Unit 2 **Using Fractions**

Apply

Solve the problems.

2 Marco is visiting the Statue of Liberty. Marco reads that the right arm of the statue is 42 feet long. That is $9\frac{1}{3}$ times the length of the statue's nose. How long is the nose?

Ask Yourself
What equation would I use if the length of the arm was 10 times the length of the nose?

Let n stand for the length of the statue's nose.

What if the length of the statue's arm was 10 times the length of its nose?

42 feet = _____ × n

Now solve the original problem.

Hint Write a sentence and an equation to solve the simpler problem first.

Answer _____

Summarize Should the answer be greater or less than 42 feet? Explain.

_____

_____

3 The Washington Monument is about 555 feet tall. That is $21\frac{1}{4}$ feet taller than $1\frac{3}{4}$ times the height of the Statue of Liberty. To the nearest foot, how tall is the Statue of Liberty?

Hint Write simpler equations showing that a 500-foot monument is 10 feet taller than twice the height of the statue.

Try simpler numbers.

500 feet − 10 feet = _____ feet

This is twice the height of the statue.

490 feet ÷ 2 = _____ feet

This is the height of the statue. Now solve the original problem.

Ask Yourself
What equation can I use to represent $1\frac{3}{4}$ times the height of the statue?

Answer _____

Apply How are the equations for this problem different from the equations you have used so far?

_____

_____

Lesson 8   **Strategy Focus: Solve a Simpler Problem**   75

Ask Yourself

How is this problem different from other problems I have solved in this lesson?

Hint Choose numbers you can use to solve a simpler problem.

④ Marcel is making a model of the face of the Crazy Horse Memorial. The face of Crazy Horse in the memorial is $87\frac{1}{2}$ feet tall. In Marcel's model, the face is $1\frac{1}{4}$ feet tall. How many times taller is the actual face than the face in the model?

The operation you should use to solve this problem is _____.

Answer _____

Choose What simpler problem did you use to help you solve this problem? How did it help?

_____

_____

⑤ In 1961, Anna Huntington sculpted a monument to Sybil Ludington, a heroine of the American Revolution. Its base is a rectangular prism. The width of the base is 4 feet 3 inches. The height is 5 feet 2 inches. The width of the base is 5 inches less than half the length of the base. What is the length of the base?

Hint You can rewrite inches as fractions of a foot.

The first operation should be _____.

The second operation should be _____.

Ask Yourself

What simpler numbers can I use to help me understand which operations to use?

Answer _____

Analyze What information given in the problem is *not* needed?

_____

_____

76 Unit 2 **Using Fractions**

Practice

Solve the problems. Show your work.

6 Toussaint L'Ouverture led a revolt in the 1790s that freed Haiti from French rule. There is a $13\frac{1}{2}$-foot monument to L'Ouverture in Miami. Pierre wants to make a model of this statue that is $\frac{1}{9}$ the size of the monument. How tall will his model be?

Answer _____

Decide How is this problem similar to Problem 5? How is it different?

_____

_____

7 The Capitol building in Washington, D.C. has a dome in its center. The circle under the dome has a circumference of about 302 feet. There are columns equally spaced around the circumference of the circle. Measured from center to center, the columns are about $8\frac{3}{8}$ feet apart along the circle. How many columns are there around the circle?

Answer _____

Formulate Explain how you wrote the equation to solve this problem.

_____

_____

Create Look back at Problem 2. Write a problem about the length of different parts of a statue. Be sure your problem can be solved using the strategy *Solve a Simpler Problem*. Solve your problem.

UNIT 2 Review

In this unit, you worked with four problem-solving strategies. You can often use more than one strategy to solve a problem. So if a strategy does not seem to be working, try a different one.

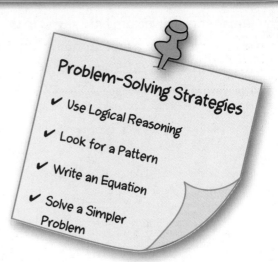

Problem-Solving Strategies
- ✓ Use Logical Reasoning
- ✓ Look for a Pattern
- ✓ Write an Equation
- ✓ Solve a Simpler Problem

Solve each problem. Show your work. Record the strategy you use.

1. You are making a box to hold your smallest drill bits. You want to arrange the bits from smallest to largest. The bits are all measured in inches. In what order will you place them?
   Bits: $\frac{1}{64}, \frac{3}{64}, \frac{5}{64}, \frac{7}{64}, \frac{9}{64}, \frac{11}{64}, \frac{1}{32}, \frac{3}{32}, \frac{5}{32}, \frac{1}{16}, \frac{1}{8}$

2. A car is 171 inches long. Paul has a model that is $\frac{1}{18}$ the size of the actual car. How long is Paul's model?

Answer _____

Strategy _____

Answer _____

Strategy _____

78 Unit 2 Using Fractions

3. Mr. Jameson is buying a new door. The door is between 2 and 3 feet wide. When written as a mixed number in simplest form, the denominator of the fraction is a multiple of 2. The sum of the numerator and denominator is 7. The numerator is one less than the denominator. What is the width of the door?

Answer _____

Strategy _____

4. Toni has a print of a famous painting. The actual painting is $92\frac{1}{10}$ centimeters tall. Her print is $30\frac{7}{10}$ centimeters tall. How many times taller is the actual painting than Toni's print?

Answer _____

Strategy _____

5. The height of the tower at 1 Houston Center is 6 meters greater than $1\frac{1}{2}$ times the height of the El Paso Tower in Houston. The tower at 1 Houston Center is 207 meters tall. What is the height of the El Paso Tower?

Answer _____

Strategy _____

Explain how you solved the problem.

Solve each problem. Show your work. Record the strategy you use.

6. A small deck is next to the front door of a house. The floor of the deck is made of two different sized planks, as shown from above. The deck is 60 inches wide. What is the length of the deck?

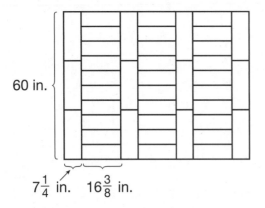

Answer _____

Strategy _____

7. Cheng catches four fish weighing $1\frac{3}{4}$ pounds, $9\frac{1}{2}$ pounds, $\frac{3}{4}$ pound, and $2\frac{5}{8}$ pounds. One of them is a Tiger Muskie. The weights of the Sunfish and the Bullhead have the same denominator. The weight of the Rainbow Trout is $3\frac{1}{2}$ times the weight of the Sunfish. What are the names of the fish in order from least to greatest weight?

Answer _____

Strategy _____

8. Mr. Kenta is paving a driveway that is $21\frac{7}{8}$ feet long. The driveway is $1\frac{3}{4}$ times as long as it is wide. What is the width of the driveway?

Answer _____

Strategy _____

Explain how you decided which operation to use to solve the problem.

80 Unit 2 **Using Fractions**

9. Mr. Franklin uses trucks to carry metal beams. He has thirty 20-foot beams that weigh $7\frac{1}{2}$ pounds per foot and seventeen 40-foot beams that weigh $17\frac{1}{4}$ pounds per foot. One truck can carry 4,000 pounds. How many trucks are needed to carry all the beams?

Answer _____

Strategy _____

10. A 20-foot metal beam weighs $14\frac{3}{4}$ pounds per foot. How many of these beams can be carried in a container that will hold 59,040 pounds?

Answer _____

Strategy _____

Write About It

Look back at Problem 9. 16,230 ÷ 4,000 = 4.0575, and 4.0575 rounds to 4. Why is 4 *not* the correct answer?

_____

_____

_____

Team Project: Design a Roof

Your team is designing a roof for a tool shed that is $10\frac{1}{2}$ feet long. You have $10,000 to spend on roofing material. You need to cover two sides of the roof with roofing material. (The sample design here of the roof shows one of the two sides shaded.) It will cost $25 for each square foot of roofing. How big will you make your roof?

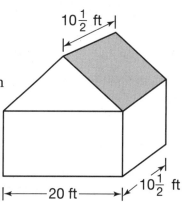

Plan
1. Discuss with your team how the spending limit will affect the size of your design.
2. Help your team find the limit on the number of square feet you can use.

Decide Agree on what size roof will be best.

Justify Make a poster showing your design, labeling the dimensions of the roof. Explain how your design fits within the budget.

Present As a group, share your design with the class. Explain how and why you made your decisions.

UNIT 3: Problem Solving Using Algebra

Unit Theme:
Explorations

People like to explore. They go deep under the sea and fly to outer space. They journey through thick jungles. Some take these trips for fun. Others explore to learn new information. In this unit, you will see how math is used in different types of explorations.

Math to Know

In this unit, you will apply these math skills:

- Add and subtract integers
- Use variables and expressions
- Solve 1- and 2-step equations

Problem-Solving Strategies

- Work Backward
- Look for a Pattern
- Write an Equation
- Solve a Simpler Problem

Link to the Theme

Write another paragraph about Shelley and Toby's tour. Include some of the facts from the table at the right.

Shelley and Toby signed up for a scuba diving tour to explore the Great Barrier Reef. These prices were listed at the sign-up booth.

Number of Dives	Price (per person)
4	$200
6	$300
8	$350

_____

_____

_____

_____

_____

82

Use Math Language

Review Vocabulary

The list below shows vocabulary terms in this unit. Knowing the meaning of these terms will help you understand the problems.

data	expression	integer	pattern
equation	function table	inverse operation	variable

Vocabulary Activity   Modifiers

A descriptive word placed in front of another word indicates a specific meaning. When learning new math vocabulary, pay attention to each word in the term.

1. A _____ has one input value for each output value.

2. You can use the information in a _____ to plot a graph.

3. An _____ reverses the effect of another operation.

4. Multiplication is the _____ of division.

Graphic Organizer   Word Map

Complete the graphic organizer.

- Write a definition of *integer*.
- Draw a diagram to show what the term means.
- Write three examples of the term.
- Write three examples of terms that are *not* integers.

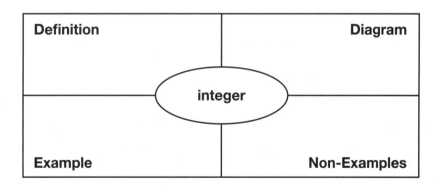

Lesson 9

Strategy Focus
Work Backward

MATH FOCUS: Integers

Learn

Read the Problem

Ocean explorers can use special vehicles (remotely-operated vehicles, or ROVs) to study the ocean floor. José is controlling an ROV that is underwater. He starts by making the ROV go down 600 feet. Then he brings the ROV back up 300 feet and moves it sideways before sending it back down 200 feet. He stops the ROV at 3,500 feet below sea level. How many feet below sea level was the ROV when José began controlling it?

Reread Tell what is happening in the problem.

- What is the problem about?
  _____

- What do you know about how the ROV moves?
  _____

- What are you asked to find?
  _____

Search for Information

Reread the problem. Decide which details will help you answer the question.

Record Write the numbers that go with those details. Write a question mark to show the number you do not know.

Start at _____ feet

Go down _____ feet

Come up _____ feet

Go down _____ feet

End at _____ feet below sea level

You know the end but you do not know the start. You can use those details to choose a problem-solving strategy.

84   Unit 3   Using Algebra

Decide What to Do

You know where José stops the ROV. You need to find the position of the ROV at the start.

Ask How can I find how many feet below sea level the ROV was when José began controlling it?

- I can use the strategy *Work Backward*.
- I can draw a diagram to show the problem situation. Then I can use inverse operations to work backward.

Use Your Ideas

Step 1 Draw and label a diagram to show what happens in the problem.

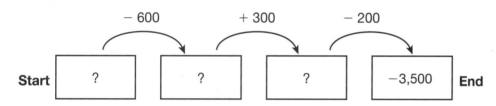

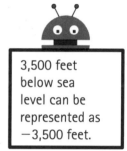

3,500 feet below sea level can be represented as −3,500 feet.

Step 2 Draw another diagram to show working backward. Use inverse operations. Record the results of each operation.

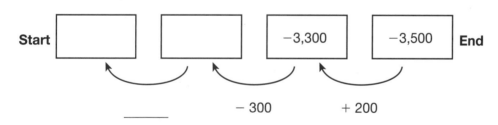

When José began controlling the ROV, it was _____ feet below sea level.

Review Your Work

Work forward through the first diagram. Check that you computed with the integers correctly.

Recognize When is working backward helpful in solving a problem?

_____

_____

Try

Solve the problem.

1. Some scientists get a grant to help pay for a deep-sea exploration. The daily costs are $20,000 for the ship, $1,000 for salaries, $200 for food, and $3,000 for the underwater robot. If the scientists can raise $75,000, they will have exactly enough money to go on a 10-day exploration. What is the amount of the grant?

Mark the Text

▪ Read the Problem and Search for Information

Identify the question you must answer. Mark information you need.

▪ Decide What to Do and Use Your Ideas

You can use the strategy *Work Backward*.

Step 1 Find the total daily cost for the exploration.

$20,000 + $1,000 + $200 + $3,000 = _____

Step 2 Find the total cost for a 10-day exploration.

$24,200 × 10 = _____

Step 3 Draw a diagram to show what happens in the problem. Use inverse operations to work backward.

Ask Yourself

If the scientists have exactly enough money for 10 days, how much money will they have left at the end of the exploration?

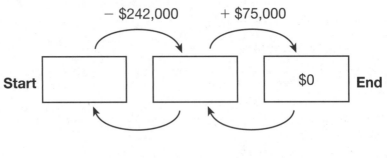

The amount of the grant is _____.

▪ Review Your Work

Check that you answered the question asked in the problem.

Identify Which phrase tells you what number to use for the end?

_____

_____

86   Unit 3   Using Algebra

Apply

Solve the problems.

(2) Researchers in the Arctic Circle recorded the average surface air temperature for 4 months. That temperature in August was 1°C warmer than in June. In October, it was 13°C cooler on average than in August. At −23°C, December's average temperature was 12°C cooler than October's. What was the surface air temperature in June?

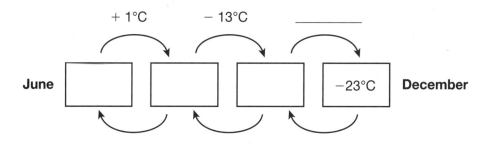

Hint Use integers to record the temperatures.

Answer _____

Explain How can you use your diagram to check your answer?

_____

Ask Yourself
How do I know what number to use for the end?

_____

(3) A scientist began observing a harbor seal already underwater. The seal dove down 325 feet, and then dove down another 250 feet. After the seal swam up 125 feet, it was 600 feet below sea level. How far below sea level was the seal after its first dive?

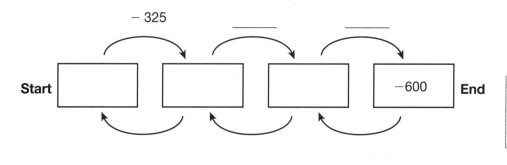

Hint Be sure that you answer the question asked.

Ask Yourself
How do I represent the location of the seal at the end?

Answer _____

Conclude Mark found the answer to be 150 feet below sea level. What mistake might he have made?

_____

Lesson 9 **Strategy Focus: Work Backward** 87

Ask Yourself

Are there some numbers in the problem that will *not* help me find the answer?

④ Lana uses the *Hercules* ROV to explore the ocean. *Hercules* is below the surface. Lana makes it dive straight down 625 meters. Next, it moves up 318 meters, east 720 meters, and then down 915 meters. *Hercules* is now at 2,500 meters below sea level. What distance from the surface was *Hercules* when Lana began controlling it?

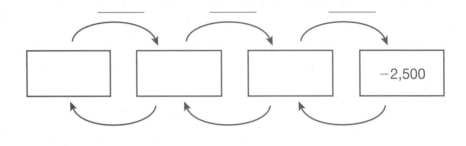

Hint Estimate: Was *Hercules* less than or more than 2,500 meters below sea level at the start?

Answer _____

Determine What information did you *not* need to solve this problem?

_____

_____

⑤ The Minerals Revenue Management Program (MRM) handles the money made from mining on government land. In 2009, MRM gave a total of about $1,498 million to three groups. The Land and Water Conservation Fund (LWCF) received some money. American Indians received $449 million, and the Historic Preservation Fund received $150 million. Did the LWCF receive more or less than $900 million? How much more or less?

Hint If MRM distributes the entire amount to three funds, the end amount is zero dollars.

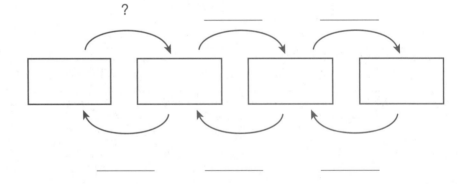

Ask Yourself

What numbers should I write in the diagram? Do they represent numbers in the ones, thousands, or millions?

Answer _____

Interpret How did you decide which operation to use?

_____

_____

88 Unit 3 **Using Algebra**

Practice

Solve the problems. Show your work.

6 The temperature of sea ice is warmer at the bottom of the ice than at its surface. The temperature at 40 centimeters below the surface of the ice is 2°C warmer than the temperature at 10 centimeters below the surface. At −5°C, the temperature at 60 centimeters below the surface is 3°C warmer than the temperature at 40 centimeters below the surface. What is the temperature of sea ice at 10 centimeters below its surface?

Answer _____

Discuss Explain how a diagram can help you solve this problem.

_____

_____

7 Jessica is a scientist controlling an ROV that can go to a depth of up to 7,200 feet. The ROV is currently below sea level. Jessica has the ROV make three stops. First, it comes up 900 feet. Then it goes down 1,400 feet. Finally, it goes down another 1,800 feet for its last stop at 6,000 feet below sea level. How far below sea level was the ROV before the three stops?

Answer _____

Modify Ask and answer another question the problem could have asked.

_____

_____

Create Look back at the problems in this lesson and choose one. Change two numbers in the problem. Then solve your new problem.

Lesson 10

Strategy Focus
Look for a Pattern

MATH FOCUS: Variables and Expressions

Learn

▪ Read the Problem

Many different satellites orbit Earth. Suppose that a satellite will make 2 orbits in 450 minutes, 3 orbits in 675 minutes, 4 orbits in 900 minutes, and 5 orbits in 1,125 minutes. How many hours will it take the satellite to make 12 orbits?

Reread Use your own words to retell the problem.

- What is the problem about?
  _____

- What kind of information is given in the problem?
  _____

- What are you asked to find?
  _____

Mark the Text →

▪ Search for Information

Read the problem again. Mark information you will need.

Record Identify information you need in the problem.

The satellite will make 2 orbits in _____ minutes.

It will make 3 orbits in _____ minutes.

It will make 4 orbits in _____ minutes.

It will make 5 orbits in _____ minutes.

This information can help you to choose a problem-solving strategy.

Decide What to Do

You know how long the satellite takes to make 2, 3, 4, and 5 orbits.

Ask How can I figure out how many hours it will take the satellite to make 12 orbits?

- I can use the strategy *Look for a Pattern*.
- I can make a table to show the information I have. Then I can write an expression for the rule and use it to find the number of hours for 12 orbits.

Use Your Ideas

Step 1 Make a table to show the information in the problem.

Orbits (n)	2	3	4	5
Time (minutes)	450	675	900	1125

Step 2 Look for a pattern in the table.

There are _____ minutes between orbits.

Step 3 Relate the time to the number of orbits, n.

2 orbits take 2 × 225 minutes.

3 orbits take 3 × _____ minutes.

4 orbits take 4 × _____ minutes.

5 orbits take 5 × _____ minutes.

The expression relating time to the number of orbits is _____ .

Step 4 Find the number of hours to make 12 orbits.

_____ × _____ minutes = _____ minutes

_____ minutes ÷ 60 = _____ hours

The satellite will take _____ hours to make 12 orbits.

Review Your Work

Check your expression by calculating the time for a different number of orbits. For example, 6 orbits should take half as long as 12 orbits.

Explain How do you know that a rule relating the time to the number of orbits involves multiplication?

_____

_____

91

Try

Solve the problem.

1 Julia is studying the Space Shuttle program. She knows the speed at which the shuttle orbits Earth. The table shows the number of hours that it took the shuttle to make a certain number of orbits. How many hours will it take the shuttle to orbit Earth 47 times?

Orbits (n)	10	11	12	13
Time (hours)	16	17.6	19.2	20.8

Mark the Text

▪ Read the Problem and Search for Information

Reread the problem and study the table. Identify the question you need to answer.

▪ Decide What to Do and Use Your Ideas

You can use the strategy *Look for a Pattern*. Look for relationships between the numbers in the table.

Ask Yourself
How does knowing the difference in the number of hours help me find a rule that relates the time to the number of orbits?

Step 1   Look for a pattern in the number of hours.
         The times are related by _____ .

Step 2   Relate the time in hours to the number of orbits, n.
         It takes _____ hours for each orbit.

Step 3   Write an expression that shows the rule for this relationship.
         An expression that shows the number of hours it takes the shuttle to orbit Earth n times is _____ .

So it will take the shuttle _____ hours to orbit Earth 47 times.

▪ Review Your Work

Try your rule. Find its value for $n = 13$. Then check to see that your result fits the pattern in the table.

Describe Why is it easier to solve the problem using the rule than continuing the pattern in the function table?

_____

_____

Apply

Solve the problems.

2 Abigail read about human space travel. She learned that extra food is provided if a landing is delayed up to 2 days. The table below shows the extra calories of food astronauts have on board. How many extra calories would 9 astronauts need for 2 days?

Astronauts (a)	4	5	6	7
Extra calories	16,800	21,000	25,200	29,400

◀ **Hint** The calories given in the table are for 2 days.

What operation relates the number of astronauts and the extra calories? _____

How many extra calories have been provided for 1 astronaut for 2 days? _____

Ask Yourself
Do I need to find out how many calories they need per day?

Answer _____

Apply How does the table help you solve the problem?

_____

_____

3 One Astronomical Unit (AU) is about 93 million miles. A space probe is traveling away from Earth. The farther it gets, the longer it takes messages to get back to Earth. The table shows approximate times for the messages over different distances. About how long will it take a message to reach Earth from a space probe 14 AU away?

Distance (d)	1 AU	2 AU	3 AU	4 AU	5 AU
Time (minutes)		16.6	24.9	33.2	41.5

Ask Yourself
How can the pattern in the times help me write an expression for the rule relating time and distance?

What operation relates the time and the distance? _____

About how long will it take a message to go 1 AU? _____

◀ **Hint** Use your rule to solve the problem.

Answer _____

Determine How did you find the pattern in the times?

_____

_____

④ *Voyager 1* spacecraft is now farther from Earth than any artificial object. The distance from Earth to the sun is 1 AU. *Voyager 1* was about 113 AU from the sun in 2010. The table shows the distance that NASA scientists expect it will travel away from where it was in 2010. About how far from the sun will *Voyager 1* be in 2020?

Year	2012	2013	2014	2015
Years after 2010				
Distance from the 2010 Location (AU)	7.2	10.8	14.4	18.0

Ask Yourself
What operation relates the distance to the number of years after 2010?

Hint The question asks how far from the sun *Voyager 1* will be in 2020, not how far it will be from its 2010 location.

▶ **Answer** _____

Identify What information is not needed to solve the problem?

_____

_____

⑤ The *International Space Station* (ISS) is the largest structure that people have built in space. The ISS orbits Earth at about 350 kilometers above Earth's surface. The ISS makes 2 orbits around Earth in about 3.06 hours, 3 orbits in 4.59 hours, 4 orbits in 6.12 hours, and 5 orbits in 7.65 hours. About how long will it take to make 21 orbits?

Orbits around Earth				
Time (hours)				

Hint Complete the table to show the information in the problem.

Ask Yourself
What operation relates the time to the number of orbits around Earth?

▶ **Answer** _____

Conclude How can you estimate to see if your answer makes sense?

_____

_____

94 Unit 3 **Using Algebra**

Practice

Solve the problems. Show your work.

6 Leslie is studying the history of space flight. The United States flew the last solo spacecraft in 1963. This was the last of the Mercury flights and lasted the longest. The table shows the times the astronaut took to make his orbits of Earth. How long did 22 orbits take?

Orbits	2	3	4	5
Time (minutes)	177	265.5	354	442.5

Answer _____

Discuss How is this problem like Problem 1? Explain.

_____

_____

7 Suppose the Hubble Space Telescope passes over your school at 8 A.M. Each passing is an orbit. The telescope will be at the same longitude 97 minutes later, then 194 minutes later, and then 291 minutes later. How many complete orbits would the telescope have made by 5 P.M.?

Answer _____

Support Jeff says the telescope will make 7 orbits between noon and midnight. Explain how your work supports this statement.

_____

_____

Create Look back at the problems and choose one. Change two of the numbers in the problem to make a new problem. Write and solve your new problem.

Lesson 10 **Strategy Focus: Look for a Pattern** 95

Lesson 11

Strategy Focus
Write an Equation

MATH FOCUS: Solve 1-Step Equations

Learn

▪ Read the Problem

> A cargo plane carries 50 passengers and many pounds of supplies to McMurdo Station, Antarctica. The total weight of the passengers is 9,000 pounds. The weight of the passengers is 2,000 pounds less than the weight of the supplies. What is the total weight of the passengers and supplies?

Reread Think of these questions as you read.

- What is the problem about?

    _____

- What is the plane carrying?

    _____

- What do you need to find?

    _____

▪ Search for Information

Mark the Text

Reread the problem. Mark information you need.

Record List the important data.

The total weight of the passengers is _____ pounds.

The weight of the passengers is _____ pounds less than the weight of the supplies.

Think about how you can use this information to solve the problem.

Decide What to Do

You know the weight of the passengers and how that weight relates to the weight of the supplies.

Ask How can I find the total weight of the passengers and supplies?

- I can use the strategy *Write an Equation*.
- I can use the weight of the passengers to find the weight of the supplies. Then I can add the weights to find the total weight.

Use Your Ideas

Step 1 Write an equation to find the weight of the supplies. Use s to represent the weight of the supplies, in pounds.

weight of supplies − weight of passengers = 2,000 pounds

s −   _____   = 2,000

Step 2 Solve the equation.

$s - 9,000 = 2,000$

$s - 9,000 + 9,000 = 2,000 + 9,000$

$s = $ _____

The supplies weighed _____ pounds.

Step 3 Add to find the total weight of passengers and supplies.

weight of passengers + weight of supplies = total weight

9,000   +   _____   =   _____

So the total weight of the passengers and supplies is _____ pounds.

> Remember to add the same number to both sides of an equation.

Review Your Work

Substitute your value of s into the equation in Step 1. Is the difference 2,000?

Explain How does writing an equation in words help you to solve the problem?

_____

_____

Try

Solve the problem.

1) The supplies for McMurdo Station in Antarctica are shipped in from other places. A tanker going to the station can carry 9 million pounds of cargo. This amount is about $\frac{3}{5}$ as much as a cargo ship can carry. About how many pounds can a cargo ship carry?

Mark the Text

▪ Read the Problem and Search for Information

Identify information you need. Make sure you understand how the numbers are related.

▪ Decide What to Do and Use Your Ideas

You can write an equation that relates the amount the cargo ship can carry to the amount the tanker can carry.

Step 1 Let c represent the amount a cargo ship can carry in millions of pounds. Write an equation to represent the situation.

$$\frac{3}{5} \times \text{cargo ship can carry} = \text{cargo tanker can carry}$$

$$\frac{3}{5} \times \underline{} = \underline{}$$

Step 2 Solve the equation.

$$\frac{3}{5}c = 9$$

$$\underline{} \times \frac{3}{5}c = \underline{} \times 9$$

$$c = \underline{}$$

Ask Yourself

What can I multiply $\frac{3}{5}$ by so that the product is 1?

The cargo ship can carry _____ million pounds of cargo.

▪ Review Your Work

Reread the problem. Check that your answer makes sense.

Recognize Why can you use the number 9 in the equation instead of 9,000,000?

_____

_____

98 Unit 3 **Using Algebra**

Apply

Solve the problems.

2 Special planes move passengers and cargo to and from Antarctica. On one flight, the total weight of 36 passengers was 7,020 pounds. What was the average weight of a passenger on the plane?

> **Hint** You can start by writing a word equation.

Let w represent the average weight of a passenger, in pounds.

number of passengers × average weight = total weight

_____ × w = _____

> **Ask Yourself**
> Which variable can I use to represent the average weight of a passenger?

Answer _____

Relate What other equation can you use to solve this problem?

_____

_____

3 Several types of penguins live in Antarctica. A chinstrap penguin weighs about 10 pounds. The difference between the weight of a chinstrap penguin and an emperor penguin is about 56 pounds. What is the approximate combined weight of a chinstrap penguin and an emperor penguin?

> **Hint** The word *difference* means to compare by subtracting.

Let e represent the weight of an emperor penguin, in pounds.

weight of emperor − weight of chinstrap = 56 pounds

e − _____ = _____

> **Ask Yourself**
> Does an emperor penguin weigh more or less than a chinstrap penguin?

Answer _____

Analyze Will uses the equation $56 + 10 = e$ to find the weight of an emperor penguin. Phil uses $e - 56 = 10$. Will they both get the correct answer? Explain.

_____

_____

Lesson 11 **Strategy Focus: Write an Equation**   99

Ask Yourself
How might I write my equation using words?

Hint Another way to write $2\frac{1}{2}$ is $\frac{5}{2}$.

④ In 1911, two teams raced to reach the South Pole first. Captain Amundsen led the Norwegian team. Captain Scott led the British team. Amundsen's team had 52 dogs. Amundsen's team had about $2\frac{1}{2}$ times as many dogs as Scott's team. About how many dogs did the teams have in all?

Let *d* represent the number of dogs Scott's team had.

_____ = _____

Answer _____

Interpret How does the *phrase $2\frac{1}{2}$ times as many dogs* help you write your equation?

_____

_____

⑤ One plane used to fly to Antarctica can carry up to 10,500 pounds of cargo and 36 passengers at a time. Suppose that in one season there was a total of 2,125 passengers on 425 flights. On average, about how many fewer passengers than the maximum number were on each flight?

Hint Use the variable *p* to represent the average number of passengers on each flight.

_____ = _____

Ask Yourself
How can I find the average number of passengers on each flight?

Answer _____

Infer Can you tell if your answer will be less than or greater than 36 before you solve the problem? Explain your answer.

_____

_____

100   Unit 3   **Using Algebra**

Practice

Solve the problems. Show your work.

6 On one flight, a plane traveled 2,400 miles to Antarctica and carried 28 passengers. This was $\frac{7}{12}$ of the maximum number of passengers. How many passengers can the plane hold?

Answer _____

Evaluate What information is given that is not needed to solve the problem?

_____

_____

7 Southern elephant seals live in Antarctica. Male elephant seals can grow to be 6 meters long and can have masses that are 8 times the mass of a female elephant seal. Suppose a male elephant seal has a mass of 3,600 kilograms. Its mass is $7\frac{1}{2}$ times the mass of a female elephant seal. How much more is the mass of the male elephant seal than the mass of the female elephant seal?

Answer _____

Adapt Suppose you know the female elephant seal's mass but not the male elephant seal's mass. How could you find the male elephant seal's mass?

_____

_____

Create Look back at the problems in the lesson. Choose one problem. Change at least two numbers. Then write and solve a new problem by writing and solving an equation.

Lesson 12

Strategy Focus
Solve a Simpler Problem

MATH FOCUS: Solve 2-Step Equations

Learn

Read the Problem

> A museum in Cincinnati gives tours of an old subway tunnel. On each tour there are 50 people. Each ticket sold raises $55 to support educational programs at the museum. If the subway tours raised $13,750 for educational programs last year, how many tours were given?

Reread Ask yourself questions while you read.

- What is the problem about?
  _____

- What kinds of information do you know about the tour?
  _____

- What question are you asked to answer?
  _____

Mark the Text →

Search for Information

Read the problem again. Mark the information you will need.

Record Write what you know about the subway tour.

How many people can go on each tour? _____

What is the cost of each ticket? _____

How much money was raised last year? _____

Think about a strategy that can help you use this information to solve the problem.

Unit 3 Using Algebra

Decide What to Do

You know how many people can go on one tour. You also know the cost of one ticket and how much money was raised last year.

Ask How can I find out how many subway tours were given last year?

- I can use the *Solve a Simpler Problem* strategy.
- I can use simpler numbers to help me figure out how to solve the problem. Once I know how to solve the problem, I can use the actual numbers to solve.

> Using simpler numbers can help you see what equation to write.

Use Your Ideas

Step 1 Use simpler numbers. Change the cost of one ticket to $10, the number of people on the tour to 5, and the money raised to $100. Let n stand for the number of tours. Then write an equation.

people per tour × ticket cost × number of tours = money raised

_____ × _____ × n = $100

n = _____

Step 2 Use the actual numbers in your equation. Then solve.

people per tour × ticket cost × number of tours = money raised

_____ × _____ × n = $13,750

_____ n = $13,750

n = _____

So _____ subway tours were given last year.

Review Your Work

Substitute 5 for n in the equation in Step 2 and solve. Did you get $13,750?

Explain Why are the numbers 5, 10, and 100 simpler numbers to use in your equation?

_____

_____

Try

Solve the problem.

1) Cave A, the deepest cave in the United States, is in Utah. It is 189 feet deeper than Cave B in Georgia. Cave C in New Mexico is 22 feet deeper than Cave B. If you add the depths of the three caves, you get 3,211 feet. What is the depth of each cave?

Mark the Text

▪ Read the Problem and Search for Information

Underline information you need. Think about how the problem relates the cave depths to one another.

▪ Decide What to Do and Use Your Ideas

You can *Solve a Simpler Problem* to find the depth of each cave.

Step 1 You can make the problem simpler by using just one variable to represent the lengths of all three caves. The cave depths are given in terms of Cave B.

Cave A is 189 feet deeper than Cave B.
Cave C is _____ feet deeper than Cave B.
Let b represent the depth, in feet, of Cave B.
So the depth of Cave C is $b + 22$.
The depth of Cave A is $b +$ _____ .

Ask Yourself

What expressions can I write to show how the depths of Cave A and Cave C are related to the depth of Cave B?

Step 2 Write and solve an equation to find the value of b.

$b +$ _____ $+$ _____ $= 3{,}211$

$b =$ _____

Step 3 Use the value of b to find the depth of each cave.

Cave B: _____

Cave A: _____ $+ 189 =$ _____

Cave C: _____ $+ 22 =$ _____

So Cave A is _____ feet deep, Cave B is _____ feet deep, and Cave C is _____ feet deep.

▪ Review Your Work

Check that the depths that you found make sense in the problem.

Clarify How does using one variable make the problem simpler?

_____

_____

Apply

Solve the problems.

2 About 125 caves in the United States are open to visitors. Some caves are publicly owned and some caves are privately owned. Of the publicly-owned caves, twice as many are in state parks as are in national parks. There are 35 more privately-owned caves than are publicly-owned. How many of the caves are privately owned?

$u + (u + \underline{}) = \underline{}$

Ask Yourself

How is the number of privately-owned caves related to the number of publicly-owned caves?

◀ **Hint** Write an equation in terms of publicly owned caves. Let u stand for the number of publicly-owned caves.

Answer _____

Identify What information is *not* necessary to solve the problem?

_____

_____

3 The largest cave room in the United States is the Big Room in Carlsbad Cavern. Its maximum length is eight times its maximum height. Suppose one wall is shaped like a rectangle with these dimensions. If its perimeter is 4,050 feet, what would the length and the height of the Big Room be?

The length is _____.

$h + 8 \times h + \underline{} + \underline{} = 4{,}050$

◀ **Hint** Let h stand for the height, in feet. Think of the length in terms of the height.

Ask Yourself

What equation can I write to show the perimeter of the wall?

Answer _____

Describe How can you draw and label a rectangle to check your answer?

_____

_____

Lesson 12 **Strategy Focus: Solve a Simpler Problem** 105

Ask Yourself
What simpler numbers can I use to solve this problem?

4 Every year in the United States, people improperly throw out used motor oil. They throw out about 18 times the amount of oil spilled in Alaska in 1989. That spill was about 257,000 barrels. To the nearest gallon, how many gallons are in each barrel if people improperly throw out about 194,000,000 gallons of used motor oil each year?

Let n represent the number of gallons in a barrel of oil.

Original problem: 18 $\times$ 257,000 $\times n =$ 194,000,000

Simpler problem: _____ $\times$ _____ $\times n =$ _____

Solve the simpler problem. Use the same steps to solve the original problem.

Hint Round your answer to the nearest gallon.

Answer _____

Sequence What steps did you use to solve the problem?

_____

_____

Ask Yourself
Do I need to find the number of each animal the scientist finds?

5 A scientist studies animals that live in Mammoth Cave, Kentucky. She finds and counts three kinds of animals. There are twice as many cave shrimp as there are cave crayfish. There are 13 fewer eyeless fish than cave shrimp. There are 37 animals in all. How many cave crayfish did the scientist find?

Hint Let c stand for the number of cave crayfish.

_____ + _____ + _____ = _____

Answer _____

Determine Suppose the scientist finds 25 more eyeless fish than cave shrimp. What expression could you write to show the number of cave shrimp the scientist finds?

_____

_____

106   Unit 3   **Using Algebra**

Practice

Solve the problems. Show your work.

6 Little brown bats often spend their winters sleeping in caves. Little brown bats can eat 600 mosquitoes per hour. At that rate, how many little brown bats will it take to eat at least 10,000 mosquitoes in one 8-hour night?

Answer _____

Discuss How do you interpret your answer if it is not a whole number?

_____

_____

7 Mammoth Cave National Park gives several different tours to its caves. The fee for the Wild Cave Tour is $48. Fourteen visitors went on the Wild Cave Tour. The fee for the Grand Avenue Tour is half the fee for the Wild Cave Tour. The park gave one of both tours one morning and received $2,544. How many visitors went on the Grand Avenue Tour?

Answer _____

Infer How do you know how much the fee for the Grand Avenue Tour is?

_____

_____

Create Look back at the problems and choose one. Change two of the numbers in the problem to make a new problem. Write and solve this new problem.

Lesson 12 **Strategy Focus: Solve a Simpler Problem** 107

UNIT 3 Review

In this unit, you worked with four problem-solving strategies. You can often use more than one strategy to solve a single problem. So if a strategy does not seem to be working, try a different one.

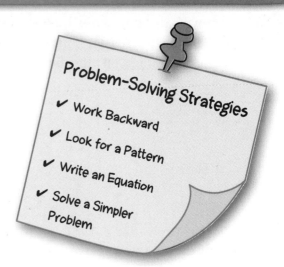

Problem-Solving Strategies
- ✔ Work Backward
- ✔ Look for a Pattern
- ✔ Write an Equation
- ✔ Solve a Simpler Problem

Solve each problem. Show your work. Record the strategy you use.

1. The temperature at the center of an iceberg is usually about $-15°C$. To convert degrees Celsius to degrees Fahrenheit, multiply degrees Celsius by $\frac{9}{5}$ and add 32. What is the temperature at the center of an iceberg when written as degrees Fahrenheit?

   Answer _____
   Strategy _____

2. ABE is a deep-sea robot. Use the table below. How much does it cost to use ABE for scientific exploration for 11 days?

Day	2	3	4	5
Cost	$7,000	$10,500	$14,000	$17,500

   Answer _____
   Strategy _____

3. Carla works at a grocery store. She works 35 hours a week and gets paid $8.50 an hour. For every hour over 35, Carla gets paid $10 an hour. If she was paid $357.50 last week, how many hours over 35 did Carla work?

Answer _____

Strategy _____

4. A space shuttle travels at about 17,500 miles per hour once it is in orbit. About how far does it travel in 3 hours 45 minutes while it is in orbit?

Answer _____

Strategy _____

5. Suppose you are in a machine that can both fly and dive. When you take over the controls, you go up 5,000 feet, and then down three times as far. You are then at −2,500 feet. When you took over, at what altitude was the machine?

Answer _____

Strategy _____

Explain how you know your answer makes sense.

Solve each problem. Show your work. Record the strategy you use.

6. The table below gives information about the rate at which an elephant's heart beats. About how many times does an elephant's heart beat in one hour?

Minutes	5	10	15	20
Number of Heart Beats	150	300	450	600

Answer _____

Strategy _____

7. The high temperature at the North Pole on a Tuesday in July was 9°C cooler than it was on Monday. On Wednesday, it was 7°C cooler than it had been on Tuesday. At 0°C, Thursday's high temperature was 15°C warmer than Wednesday's high temperature. What was the high temperature on Monday at the North Pole?

Answer _____

Strategy _____

8. Think of the ocean's surface as altitude 0. The top of an iceberg off the coast of Antarctica is 165 feet above the surface. The bottom of the same iceberg is −1,155 feet below the surface. What is the total length of the iceberg?

Answer _____

Strategy _____

Explain how you can solve this problem without adding negative numbers.

110   Unit 3   Using Algebra

9. A roughly rectangular ice sheet in Alaska has a volume of about 1,925,000 cubic meters. This ice sheet is about 125 meters wide and 6.16 meters thick. About how long is it?

Answer _____

Strategy _____

10. You are keeping track of your family's drive to the campground. The campground is about 330 miles from home. If you do not stop before you get there, about what time will you arrive?

Time	8 A.M.	9 A.M.	10 A.M.	11 A.M.
Distance Traveled (miles)	0	55	110	165

Answer _____

Strategy _____

Write About It

Look back at Problem 9. Describe how you used the information in the problem to choose a strategy for solving it.

_____

_____

_____

Team Project: Plan an Adventure

The 27 students in your class want to explore one of two caves for a field trip. The budget for the field trip is $700. Seven adults will go on the trip as drivers. Information about both caves is shown at the right.

Plan
1. The cost for a trip will include admission, one tour, and lunch for everyone in the group. Make a table to show the costs for four possible field trips to the caves.
2. Compare the costs to the $700 in the budget for the field trip. Identify the trips that the class can take while staying within the budget.
3. Think about how to schedule activities for each trip.

Decide Choose which trip your team will recommend for the class field trip.

Create Make a poster of the activities for the field trip.

Present As a group, share your plan with the class. Discuss the reasons for your choice.

Deep Cave
Adult admission: $10
Student admission: $7
1-hour tour: $6
3-hour tour: $15
Picnic lunch: $5

Long Cave
Adult admission: $7
Student admission: $5
1-hour boat tour: $6
2-hour boat tour: $11
Box lunch: $4

Unit 3 Review 111

UNIT 4: Problem Solving Using Ratio, Proportion, Percent, and Probability

Unit Theme:
At the Mall

When you go shopping, you may be going to a market, a department store, or a cart on the street corner. Shopping is about more than just going to the store. It is also about the people who work there and the people who own the businesses. In this unit, you will see how math is used in shopping.

Math to Know

In this unit, you will apply these math skills:

- Work with ratios and develop proportional thinking
- Solve problems with percents
- Find the probability of compound events

Problem-Solving Strategies

- Make a Table
- Use Logical Reasoning
- Write an Equation
- Make an Organized List

Link to the Theme

Write another paragraph about the new store. Which coupon will Betsy receive? What will she buy with it?

A new store is having its grand opening. Betsy heard that the first 100 customers will receive 50%-off coupons from the store. The second 100 customers will receive 25%-off coupons. The third 100 customers will receive 10%-off coupons.

_____
_____
_____
_____
_____

Use Math Language

Review Vocabulary

The list below shows vocabulary terms in this unit. Knowing the meaning of these terms will help you understand the problems.

cross multiply   favorable outcome   probability   ratio
discount   outcome   proportion   tree diagram
equivalent ratios percent   rate

Vocabulary Activity   Math Terms

Some terms are found only in math. Use terms from the list above to complete the following sentences.

1. The _____ of boys to girls in the room is 2:3.

2. 1:4 and 2:8 are _____ because they can be represented by equivalent fractions.

3. 25 _____ is another way to say 1 ÷ 4.

4. When you _____ , you multiply one numerator and the opposite denominator in a pair of equivalent ratios.

Graphic Organizer   Word Web

Complete the graphic organizer.

- Write a definition of the term *ratio*.
- Find three terms from the vocabulary list that have something to do with a ratio.
- Write a definition of each of the three terms.

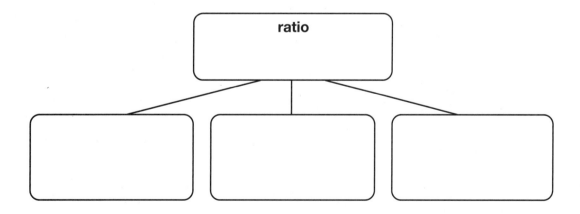

113

Lesson 13

Strategy Focus
Make a Table

MATH FOCUS: Rates and Ratios

Learn

▢ Read the Problem

> The Reyes family is going to the mall. When they leave home, the car odometer reads 54,362 miles. When they arrive at the mall 15 minutes later, the odometer reads 54,372 miles. What was their average speed in miles per hour?

Reread Ask yourself questions as you read.

- What is the problem about?
  _____

- What information is given?
  _____

- What do you need to find?
  _____

Mark the Text ⟶

▢ Search for Information

Read the problem again. Look up any words you may not understand.

Record Write the words and numbers you need to solve the problem.

An odometer measures the _____ .

The starting odometer reading was _____ miles.

The ending odometer reading was _____ miles.

It took _____ minutes to get to the mall.

Think about how you can use this information to choose a problem-solving strategy.

114 Unit 4 **Using Ratio, Proportion, Percent, and Probability**

Decide What to Do

You can find the number of miles the family traveled from the odometer readings. You know how long the trip took in minutes.

Ask How can I find the average speed in miles per hour?

- First, I can find how far the Reyes traveled in 15 minutes.
- Then I can use the strategy *Make a Table* to find how far the family would travel in 60 minutes.

Average speed can be written as the ratio $\frac{distance}{time}$.

Use Your Ideas

Step 1 Find how far the Reyes traveled in 15 minutes.

Subtract the starting odometer reading from the final reading.

$54{,}372 - 54{,}362 = $ _____

At this speed, the car will travel _____ miles in 15 minutes.

Step 2 Find how far the Reyes would travel in 60 minutes.

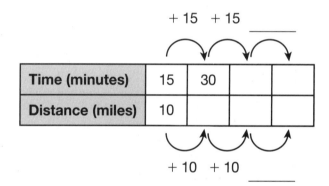

The Reyes would travel _____ miles in 60 _____.

Step 3 Write the rate in miles per hour.

$$\frac{_____ \text{ miles}}{\text{minutes}} = \frac{_____ \text{ miles}}{\text{hour}}$$

So the average speed was _____ miles per hour.

Review Your Work

Reread the problem. Check that you answered the question asked.

Describe How did the table help you solve the problem?

_____

_____

Try

Solve the problem.

1) A mall did a survey of grandmothers who shop with grandchildren. That day, 35 out of every 50 grandmothers bought lunch for their grandchildren. In all, 175 grandmothers bought lunch for their grandchildren. About how many grandmothers shopped with grandchildren that day?

Mark the Text

■ Read the Problem and Search for Information

Reread the problem and circle the important information. Underline the question.

■ Decide What to Do and Use Your Ideas

You can *Make a Table* to find equivalent ratios.

Step 1 Write the ratio in words. Then write it using numbers.

grandmothers who buy lunch : all grandmothers

35: _____

Ask Yourself

Why do I stop at 175?

Step 2 Make a table. Continue the pattern until you find where the number of grandmothers who bought lunch is 175.

Grandmothers Who Buy Lunch	35	70			
All Grandmothers	50	100	150		
Ratio	35:50	70:100			

So about _____ grandmothers shopped with grandchildren that day.

■ Review Your Work

Check that $\frac{35}{50}$ is equivalent to the ratio in the last column.

Identify How did you complete the table?

_____

_____

Apply

Solve the problems.

2) Mrs. Singh manages the Center City Mall. She knows that about 80 out of 125 teenagers go to small shops instead of big department stores. She expects about 625 teenagers will come to the mall this week. About how many of these teenagers will go to small shops instead of department stores?

Teenagers Who Visit Small Shops	80	160			
Total Teenagers	125	250	375		
Ratio	80:125	160:250			

◄ **Hint** Make a table to find equivalent ratios.

Ask Yourself
What numbers should I use in the table?

Answer _____

Employ If 500 teenagers visit the mall, how would you use the table to find about how many would visit department stores?

_____

_____

3) The managers of several malls surveyed people who went to their movie theaters. They found that the ratio of *spending by moviegoers* to *spending by non-moviegoers* is about 7 to 20. Suppose non-moviegoers spend an average of $100 each. How much more do non-moviegoers spend on average than moviegoers?

Ask Yourself
What numbers do I need to solve the problem?

Spending by Moviegoers	$7	$14			
Spending by Non-moviegoers	$20	$40			

◄ **Hint** Make sure you answer the question the problem asks.

Answer _____

Clarify Corey found the answer to be $35. What mistake could he have made?

_____

_____

Lesson 13 **Strategy Focus: Make a Table**

Ask Yourself
How can I decide how many gallons of gas are in the tank?

④ Mr. Nolan is going to a mall 35 miles away. His car's gas tank holds 12 gallons of gas. The car gets about 28 miles per gallon. The gauge indicates that the tank is $\frac{1}{4}$ full. How many miles can Mr. Nolan drive with the gas in the tank? Can he make it to the mall and back without getting gas?

Gallons			
Miles			

Hint Remember: Mr. Nolan needs to get to the mall and back.

▶ **Answer** _____

Explain How did you use the table to answer the question?

_____

_____

Ask Yourself
How can I write 52% as a ratio?

⑤ The owners of a mall want to add movie theaters. They expect that an average of 150 people will go to each theater every day. They think that about 52% of the people who go to the movies will also visit mall stores. If they put in two theaters, about how many moviegoers would they expect to visit mall stores each day?

Hint Find the total number of moviegoers at the two theaters.

Number of Mall Visitors			
Number of Moviegoers			

Answer _____

Analyze Suppose the mall owners want to put in four theaters. Without extending the table, how could you find the number of moviegoers expected to visit mall stores?

_____

_____

118   Unit 4   Using Ratio, Proportion, Percent, and Probability

Practice

Solve the problems. Show your work.

6 Mr. Edwards is going shopping for new pants and shirts. The sales tax where he lives is 6.25%. He gets a $2-off coupon for every $20 he spends at the mall. How much would Mr. Edwards need to spend to get $12 off in coupons?

Answer _____

Conclude What information is given that is not needed to solve the problem?

_____

_____

7 Mrs. Ling rents space at the mall where she sells sunglasses. She must pay the mall $50 every day, plus $6 for every $125 in sales. Today, she paid the mall $92. What were her sales today?

Answer _____

Interpret Can you use the ratio $\frac{50}{125}$ to solve this problem? Why or why not?

_____

_____

Create A car can drive 25 miles per gallon of gas. Use this rate to write a problem that can be solved using the strategy *Make a Table*. Solve your problem.

Lesson 13 **Strategy Focus: Make a Table** 119

Lesson 14

Strategy Focus
Use Logical Reasoning

MATH FOCUS: Equivalent Ratios

Learn

Read the Problem

Mrs. Dean wants to buy a cart to set up in the mall. A cart with an area of 36 square feet costs $7,900. If she can afford it, Mrs. Dean would rather have a cart with an area of 64 square feet. If the ratio of cost to floor area stays the same, how much would she have to pay for a cart with an area of 64 square feet? Round to the nearest dollar.

Reread Ask yourself questions about the problem as you read.

- What does Mrs. Dean want to do?
  _____

- For what size cart does she know the ratio of cost to floor area?
  _____

- What question am I asked to answer?
  _____

Mark the Text → Search for Information

Read the problem again. Find the facts that you need.

Record Write the key information from the problem.

The given ratio of cost to floor area is _____.

The floor area of the cart Mrs. Dean wants is _____.

Think about how you can use these facts to solve the problem.

Decide What to Do

You know the ratio of cost to floor area for a small cart. You also know the floor area of the large cart.

Ask How can I find the cost of the large cart?

- The ratio I know is $\frac{\$7{,}900}{36 \text{ square feet}}$. I can set up a proportion by writing an equivalent ratio using 64 square feet.

- I can use the strategy *Use Logical Reasoning* to help me set up the proportion.

Use Your Ideas

Step 1 Write the ratio in words. Then fill in what you know.

$$\frac{\text{cost}}{\text{floor area}} = \underline{}$$

Step 2 Write what you know in the equivalent ratio.

$$\frac{\$7{,}900}{36 \text{ square feet}} = \frac{\text{cost}}{\underline{}}$$

Step 3 You can cross multiply to find the missing numbers.

$$\frac{7{,}900}{\underline{}} = \frac{\text{cost}}{\underline{}}$$

$7{,}900 \times \underline{} = \underline{} \times \text{cost}$ ← Cross multiply.

$505{,}600 = 36 \times \text{cost}$ ← Simplify.

$\frac{505{,}600}{36} = \frac{36 \times \text{cost}}{36}$ ← Divide.

$\underline{} \approx \text{cost}$ ← Round.

If *cost* is in the numerator in one ratio, *cost* should be in the numerator of the equivalent ratio.

So Mrs. Dean would pay _____ for a cart with an area of 64 square feet.

Review Your Work

Decide whether your answer makes sense. Thirty-six is more than half of 64. Is 7,900 more than half of 14,044?

Recognize Could you have set up the two ratios as $\frac{\text{floor area}}{\text{cost}}$? Explain why or why not.

_____

_____

Try

Solve the problem.

1 Joel sells fruit smoothies for $3.95 at his mall cart. His recipe uses 9 ounces of yogurt for every 3 ounces of fruit. He has 57 pounds of yogurt and 10 pounds of fruit. How much more fruit does Joel need to use all the yogurt for making smoothies?

Mark the Text

■ Read the Problem and Search for Information

Circle the important information. See how the numbers are related.

■ Decide What to Do and Use Your Ideas

Use logical reasoning to help set up your proportion.

Ask Yourself

Does it matter that the first ratio has the ingredients measured in ounces and the second ratio has them measured in pounds?

Step 1 Find out how much fruit Joel needs. Write equivalent ratios. Let f represent the unknown amount of fruit.

$$\frac{\text{yogurt}}{\text{fruit}} \rightarrow \frac{9 \text{ oz}}{} = \frac{}{f}$$

Step 2 Find the value of f.

$$\frac{9}{} = \frac{}{f}$$

$9 \times \underline{} = \underline{} \times \underline{}$ ← Cross multiply.

$\underline{} = \underline{}$ ← Simplify.

$\underline{} = \underline{}$ ← Divide.

Joel needs _____ pounds of fruit to make the smoothies.

Step 3 Subtract to find the amount of fruit Joel needs.

_____ pounds − 10 pounds = _____ pounds

So Joel needs _____ more pounds of fruit.

■ Review Your Work

Make sure you answered the question asked.

Distinguish What information is not needed to solve the problem?

_____

_____

122   Unit 4   Using Ratio, Proportion, Percent, and Probability

Apply

Solve the problems.

2 Mr. Moore hires Reba to work at his mall cart. The first week, Mr. Moore pays her $200 for working 5 days. At this rate, how much will Mr. Moore pay Reba for working 44 days?

$\frac{\text{dollars}}{\text{days}} \rightarrow \frac{}{5} = \frac{p}{}$

Ask Yourself

What proportion will I write to solve the problem?

◀ **Hint** Use p to represent the amount Reba will be paid.

Answer _____

Identify What is another question you could ask from the information in the problem?

_____

_____

3 Ming paints silk scarves. She can paint 15 scarves in two hours. She sells her scarves Mondays through Fridays from a mall cart. If she sells about 18 scarves each day, about how much time does Ming need to paint the scarves she sells each week?

scarves sold per week = _____

Use h to represent the number of hours she needs to paint scarves.

$\frac{\text{scarves}}{\text{hours}} \rightarrow \frac{15}{} = \frac{}{h}$

Ask Yourself

How can I find the number of scarves Ming sells in one week?

◀ **Hint** The ratio *15 scarves* to *2 hours* must equal the ratio *number of scarves to make* to *hours to make them*.

Answer _____

Generalize How does logical reasoning help you set up your proportion?

_____

_____

Lesson 14 **Strategy Focus: Use Logical Reasoning** 123

Ask Yourself

Do I need all of the information given in the problem?

Hint There is more than one step in this problem.

④ Mr. Peters keeps track of sales and customers at his mall cart. During the first 7 days of June, he had sales of $3,000 and 348 customers. He wants to make a total of $13,000 in June. At this rate, about how many more customers does Mr. Peters need to meet his goal?

Let c represent the number of customers he needs.

$$\frac{\text{customers}}{\text{dollars}} \rightarrow \frac{}{} = \frac{}{}$$

Answer _____

Examine Colette says that Mr. Peters needs about 1,508 more customers. What mistake might she have made?

_____

_____

Hint There are 31 days in August.

⑤ Cart rentals at Mall of America are about $2,300 per month or 15% of monthly sales, whichever is greater. Mr. Gomes rents a cart from this mall. His weekly sales in August are usually about $5,320. What should Mr. Gomes expect to pay for cart rental during August?

Let x represent his expected monthly sales.

Ask Yourself

If I know the ratio of sales to 7 days, how can I find an equivalent ratio of sales to 31 days?

$$\frac{\text{dollars}}{\text{days}} \rightarrow \frac{}{} = \frac{}{}$$

$15\% \times \underline{} = \underline{}$

Answer _____

Explain How did you use the amount $2,300 to solve the problem?

_____

_____

Practice

Solve the problems. Show your work.

6 Mrs. Rose runs a cart at the mall that sells cell phones. On average, for every 4 hours the cart is open, 10 phones are sold. Her goal is to sell 175 phones each week. How many hours per week does she need to keep the cart open to meet that goal?

Answer _____

Determine A student used the ratio $\frac{10}{4}$ to solve the problem. Another student used the ratio $\frac{5}{2}$. Will both ratios work? Explain.

_____
_____

7 Mr. Gregg owns 3 pizza carts. He earns a total of about $2,775 per week from all 3 carts. Mr. Gregg wants to buy 2 more pizza carts. About how much money can he expect to earn per week from the 2 carts?

Answer _____

Relate What proportion could you use to find out how much Mr. Gregg can expect to make per week from all 5 pizza carts?

_____
_____

Create Look back at the problems in the lesson. Write and solve a problem about earnings from a mall cart that involves proportions. Make sure your problem can be solved using the *Use Logical Reasoning* strategy.

Lesson 14 **Strategy Focus: Use Logical Reasoning**   125

Lesson 15

Strategy Focus
Write an Equation

MATH FOCUS: Percent Concepts and Applications

Learn

▢ Read the Problem

> Mr. Garcia wants to buy a new pair of jeans. At Store A, the jeans he likes usually cost $54. That store has them on sale for $42. At Store B, the same jeans usually cost $52, but they are on sale for 10% off. Which store has the better deal?

Reread Ask yourself questions as you read.

- What is the problem about?
  _____

- What kind of information is given in the problem?
  _____

- What do I need to find out?
  _____

Mark the Text →

▢ Search for Information

Read the problem again. Think about the words and numbers you will use.

Record Write what you know about the prices of the jeans.

The regular price at Store A is _____ .

The sale price at Store A is _____ .

The regular price at Store B is _____ .

The discount at Store B is _____ .

You will need to use all of this information to solve the problem.

126 Unit 4 **Using Ratio, Proportion, Percent, and Probability**

Decide What to Do

You know the regular price of jeans at two stores. You know the sale price at one store and the percent of discount at the other.

Ask How can I find out if 10% off $52 is a better deal than $42?

- I can use the strategy *Write an Equation*.
- I can set up a proportion to find the discount amount, d, at Store B. Next I can find the sale price. Then I can compare the prices.

Use Your Ideas

Step 1 Write a proportion.

Write 10% as a ratio: $\frac{10}{100}$.

The ratio relates a part, 10, to the whole, 100.

Write a ratio that compares the amount of the discount, d, and the whole price at Store B, $52. Then write a proportion by setting that ratio equal to the ratio for the percent.

$$\frac{d}{52} = \frac{10}{100}$$

Step 2 Cross multiply. Solve the equation for d.

$d \times$ _____ = _____ $\times\ 10$

_____ = _____

_____ = _____

The discount amount is _____ .

Step 3 Subtract to find the sale price of the jeans at Store B.

$52.00 − _____ = _____

The jeans at Store B are on sale for _____ .

Step 4 Compare the two sale prices.

$46.80 ◯ $42.00

So _____ has the better deal.

> The discount amount is not the same as the discounted, or sale, price.

Review Your Work

Check that you compared the sale prices of the jeans at both stores.

Show How did writing an equation help you solve the problem?

_____

_____

127

Try

Solve the problem.

(1) Brianna is shopping at the mall. Her favorite T-shirts are on sale for 20% off the regular price of $15. How much will she save if she buys six T-shirts on sale?

■ Read the Problem and Search for Information

Restate the problem in your own words. Think about how to start solving the problem.

What proportion can I write that will help me solve the problem?

■ Decide What to Do and Use Your Ideas

You can use the strategy *Write an Equation* to find the discount amount on 1 T-shirt and then the total savings on 6 T-shirts.

Step 1 Set up a proportion.
Write a ratio to show the percent. Write another ratio to compare the amount of the discount, *d*, and the regular price.

$$\frac{\text{part}}{\text{whole}} = \frac{\text{part}}{\text{whole}}$$

$$\frac{d}{15} = \frac{20}{100}$$

Step 2 Cross multiply to solve for *d*.

_____ = 15 × _____

_____ = _____

_____ = _____

So the discount amount is _____ .

Step 3 Find how much Brianna will save if she buys six T-shirts.

6 × _____ = _____

If Brianna buys six T-shirts on sale, she will save _____ .

■ Review Your Work

Make sure you answered the question in the problem.

Contrast Art says he found the regular price for 6 T-shirts and then found 20% of that. His answer is correct. Why?

_____

_____

Apply

Solve the problems.

(2) Mr. Rosen is planning to buy a pair of shoes that costs $48 at the mall. The next day, he sees that those shoes are on sale for $30. What is the percent discount?

Let *p* stand for the percent of discount. Use the amount of the discount in your proportion.

Amount of discount = $48 − _____ = _____

$$\frac{\text{part}}{\text{whole}} = \frac{\text{part}}{\text{whole}}$$

$$\frac{p}{100} = \underline{\qquad}$$

> **Ask Yourself**
> What ratio can I write to find the unknown percent?

> ◀ **Hint** Think of $48 as the whole amount and $18 as the part.

Answer _____

Examine Bettina says she knows the discount is less than 50% because 50% of $48 is $24. Is she right?

_____

_____

(3) The grocery store is having a sale on Jamie's favorite cereal. The regular price is $5.25. The sale price is $4.20. What is the percent discount?

$$\text{percent discount} = \frac{\text{regular price} - \text{sale price}}{\text{regular price}}$$

$$\frac{p}{100} = \frac{\underline{\qquad} - \underline{\qquad}}{\underline{\qquad}}$$

> ◀ **Hint** You can write everything in one proportion.

> **Ask Yourself**
> I know $5.25 is the whole. What is the part in this ratio?

Answer _____

Identify What equation would you write to find what percent of the regular price the sale price is?

_____

_____

Lesson 15 **Strategy Focus: Write an Equation** 129

④ At a 40%-off sale, Marne bought a sweater for $75. How much did Marne save?

Hint 40% off means Marne paid 60% of the regular price.

$$\frac{\text{part}}{\text{whole}} = \frac{\text{part}}{\text{whole}}$$

$$\frac{60}{100} = \frac{75}{x}$$

Ask Yourself
Once I find the regular price, how can I find the amount Marne saved?

Answer _____

Analyze Suppose the store takes an additional 40% off the sale price of the same sweater. Would the additional savings be $50? Explain.

_____

_____

Ask Yourself
How will I write 6% as a ratio?

⑤ In Rina's state, she must pay 6% sales tax on clothing. She plans to buy a shirt that costs $32. If she drives an hour to the next state to buy the same shirt for $32, she will pay no sales tax. Gas for the drive will cost $10. Will Rina save or lose money if she buys the shirt in the next state? How much money will she save or lose?

Cost of the shirt in Rina's state: _____

Hint First, find how much Rina would pay in her home state. Then find the total cost of the shirt in the next state, including the cost of the gas.

Cost of the shirt in the next state: _____

Answer _____

Demonstrate If Rina wanted to buy ten of those shirts, would she save money by buying the shirts in the next state? Explain why or why not.

_____

_____

130   Unit 4   Using Ratio, Proportion, Percent, and Probability

Practice

Solve the problems. Show your work.

6 A local sporting goods store is having a "Buy One, Get One for 50%!" sale. Mr. Connors is going to buy tennis rackets for his two grandchildren. The rackets usually cost $125 each. How much will Mr. Connors save?

Answer _____

Justify Tell why you chose the equation you used to find your answer.

_____

_____

7 In the United States, there is a federal tax on each gallon of gasoline sold. In 2010, the tax was about 18¢ per gallon. If gas sells for $2.75 per gallon, what percent of that price is federal tax? Round your answer to the nearest tenth of a percent.

Answer _____

Infer How did you use an equation to solve this problem?

_____

_____

Create Write and solve a problem about percents. Make sure that your problem can be solved using the strategy *Write an Equation*.

Lesson 15 **Strategy Focus: Write an Equation** 131

Lesson 16

Strategy Focus
Make an Organized List

MATH FOCUS: Probability

Learn

Read the Problem

> At a food court, the cashier puts one packet each of mustard, ketchup, and relish in Rhonda's bag. Rhonda wants only mustard and relish. She reaches into the bag and chooses a packet at random. Then she reaches in and chooses another packet at random. What is the probability that she will choose the packets she wants?

Reread Ask yourself these questions as you read.

- What is Rhonda doing?
  _____

- What information is given?
  _____

- What do I need to find?
  _____

Mark the Text

Search for Information

Read the problem again. There are no numbers in this problem, but there is information you need.

Record Write the information you know.

Types of packets: _____ , _____ , _____

Packets Rhonda wants: _____ and _____

Think about how this information will help you solve the problem.

132   Unit 4   **Using Ratio, Proportion, Percent, and Probability**

Decide What to Do

You know that there are 3 packets in the bag. You know that once Rhonda has chosen 1 packet, there are only 2 packets left.

Ask How can I find the probability that Rhonda chooses mustard and relish?

- I can use the strategy *Make an Organized List*. I can make a list of all of the possible outcomes.

- I can count all the possible outcomes and the outcomes that contain mustard and relish. Then I can find the probability.

Use Your Ideas

Step 1 List all the ways Rhonda can choose two packets. Start by listing the possible outcomes if she chooses mustard first. Keep choices in order so you do not miss or repeat any.

First Choice	Mustard	Mustard	Ketchup	Ketchup	Relish	
Second Choice	Ketchup	Relish	Mustard	Relish		

Step 2 Count the number of possible outcomes. Then count the number of favorable outcomes.

Number of ways to choose 2 packets: _____
Number of ways to choose mustard and relish: _____

Step 3 Write the probability as a fraction. Simplify your answer.

$$\frac{\text{number of favorable outcomes}}{\text{number of possible outcomes}} = \underline{\qquad}$$

So the probability Rhonda chooses mustard and relish is _____ .

The order in which Rhonda chooses the mustard and relish does not matter.

Review Your Work

Make sure that you listed all of the possible outcomes.

Identify How did making an organized list help you solve the problem?

_____

_____

Try

Solve the problem.

1. Rich is buying lunch. He will choose a soup and a sandwich from the menu. If all the choices are equally likely, what is the probability Rich chooses chicken soup and a ham or tuna sandwich?

> **Menu Choices**
> **Soups:** Chicken, Vegetable
> **Sandwiches:** Ham, Turkey, Tuna

▢ Read the Problem and Search for Information

Restate the problem in your own words. Think about the possible outcomes.

Ask Yourself

What are the possible outcomes? What are the favorable outcomes?

▢ Decide What to Do and Use Your Ideas

You can make an organized list to show all the possible outcomes. A tree diagram is one way to make an organized list.

Step 1 Make a tree diagram to list all of the possible outcomes.

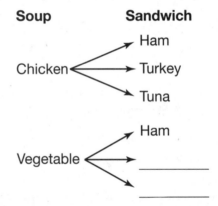

Step 2 Count the number of possible outcomes and the number of favorable outcomes. Write the probability in simplest form.

$$\frac{\text{number of favorable outcomes}}{\text{number of possible outcomes}} = \underline{}$$

So the probability that Rich chooses chicken soup and a ham or tuna sandwich is _____ .

▢ Review Your Work

Make sure that you counted every possible outcome.

(Recognize) How does a tree diagram help make an organized list?

_____

_____

134 Unit 4 **Using Ratio, Proportion, Percent, and Probability**

Apply

Solve the problems.

2) Some sixth graders go on a field trip. Half of the group is boys. For lunch, half of the students go to the sandwich shop. The others go to the pizza shop. If each student is equally likely to go to either shop, what is the probability that a student chosen at random will be a girl who went to the pizza shop for lunch?

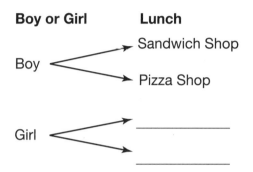

▸ **Hint** There are two categories you must consider.

Ask Yourself
How can I organize what I know in a tree diagram?

Answer _____

Clarify Why do you not need to know the number of students?

_____

_____

3) At a smoothie shop, a wheel is divided into 4 equal sections for different flavors. Customers who spin the wheel and land on the same flavor twice in a row get a free smoothie. What is the probability that a customer lands on Banana twice in a row?

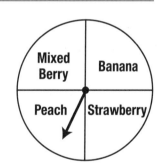

Ask Yourself
Does the outcome of the first spin affect the outcome of the second spin?

List the possible outcomes.

MM	MB	MS	MP
BM	BB	BS	___
SM	SB	___	___
___	___	___	___

◂ **Hint** Use letters to represent the flavors on your list. Use "M" for mixed berry, "B" for banana, "S" for strawberry, and "P" for peach.

Answer _____

Conclude If customers have to spin the same flavor three times to win, are they more or less likely to win a free smoothie? Why?

_____

_____

Lesson 16 **Strategy Focus: Make an Organized List** 135

Ask Yourself

Does the order in which she chooses the fork and knife matter?

④ Kaya needs a fork and a knife. There is a bin of plastic forks, a bin of plastic knives, and a bin of plastic spoons. She reaches into a bin without looking and takes an item. Then she reaches into another bin and takes another item. What is the probability she chooses a fork and knife?

Hint She chooses from a different bin the second time.

First Choice						
Second Choice						

Answer _____

Examine How would your list be different if Kaya could choose from the same bin both times?

_____

_____

⑤ A juice bar gives away free samples of orange, apple, cranberry, and grapefruit juice. Each customer has an equal chance of getting any one of the four flavors. Every tenth customer will get a coupon for a free drink. What is the probability that the first customer gets a sample of orange juice and the second does not get a sample of apple juice?

Hint Make a tree diagram to show the possible ways the samples can be given out.

1st Customer

2nd Customer

Ask Yourself

Can two people in a row get the same flavor?

Answer _____

Explain What information is not needed to answer the question?

_____

_____

136   Unit 4   Using Ratio, Proportion, Percent, and Probability

Practice

Solve the problems. Show your work.

6 Mitch and Helen are each going to lunch. They will each choose between a Chinese restaurant, an Indian restaurant, and a Mexican restaurant. They may choose to go to the same place or to different places. If all choices are equally likely, what is the probability that neither goes to the Chinese restaurant?

Answer _____

Determine Otis says the probability is $\frac{1}{3}$. What mistake might he have made?

_____

_____

7 A customer can spin one wheel to choose an item and another to find the discount. What is the probability that a customer will get $1 off a sandwich?

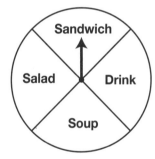

Answer _____

Investigate Ask another question about the problem situation.

_____

_____

Create Look back at the problems in the lesson. Write and solve a probability problem that can be solved by making an organized list.

Lesson 16 **Strategy Focus: Make an Organized List** 137

Unit 4 Review

In this unit, you worked with four problem-solving strategies. You can often use more than one strategy to solve a problem. If a strategy does not seem to be working, try a different one.

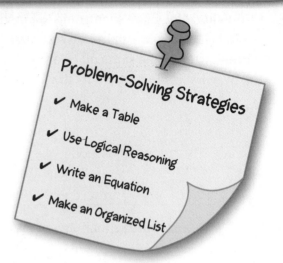

Problem-Solving Strategies
- ✔ Make a Table
- ✔ Use Logical Reasoning
- ✔ Write an Equation
- ✔ Make an Organized List

Solve each problem. Show your work. Record the strategy you use.

1. The speed limit on the road Nana takes to get to the mall is 35 miles per hour. If she drives at the speed limit for the entire trip, how many minutes will it take her to get to the mall 14 miles away?

   Answer _____

   Strategy _____

2. A store is having a 20%-off sale today only. Tomorrow, the prices will go up 20% from today's sale price. If a T-Shirt cost $15 before the discount yesterday, what will it cost tomorrow?

   Answer _____

   Strategy _____

3. At a juice bar, a special fruit drink is made with 4 ounces of orange juice and 6 ounces of grape juice. The manager has 12 gallons of orange juice. How much grape juice will the manager need if she plans to use all the orange juice?

Answer _____

Strategy _____

4. There are an equal number of red, yellow, blue, and green cubes in a paper bag. Without looking, May will choose a cube and then place it back in the bag. Then Jim will do the same. What is the probability that they will both choose a green cube?

Answer _____

Strategy _____

5. Bo is playing a game to win a free pair of sunglasses. He must estimate the number of steps it will take him to walk from his front door to the eyewear shop on his street. Bo usually walks about 10 feet for every four steps. The distance is about 225 feet. About how many steps will Bo take?

Answer _____

Strategy _____

Explain how you found the number of steps for 225 feet.

Solve each problem. Show your work. Record the strategy you use.

6. In their parking lots, many malls allow $3\frac{1}{2}$ parking spaces for every 1,000 square feet of retail space. A company is planning a mall with 2 million square feet of retail space. How many parking spaces should the group plan?

   Answer _____

   Strategy _____

7. Sal and Roger are playing a game. They roll a number cube labeled 1 through 6. The player who rolls the greater number wins. If they roll the same number, they tie. Sal rolls, and then Roger rolls. What is the probability that Roger wins?

   Answer _____

   Strategy _____

8. Lila is standing at the mall entrance to hand out coupons to four different stores. It is equally likely that a person will get a coupon for any one of the stores. Lila gives a coupon to both Rory and Rita. What is the probability they got a coupon for the same store?

   Answer _____

   Strategy _____

   Explain how you solved the problem.

9. A delivery truck has a daily route that is 60 miles long. The truck gets about 12 miles per gallon of gasoline. Its gas tank holds $17\frac{1}{2}$ gallons. About how often will the tank need to be refilled?

Answer _____

Strategy _____

10. The management of a large mall wants to increase the number of shops in the food court by 25%. There are 16 shops in the food court now. How many shops will be in the food court after the increase?

Answer _____

Strategy _____

Write About It

Look back at Problem 10. Describe how you used the information in the problem to choose a strategy for solving it.

_____

_____

_____

Team Project: Make a Mall

You are on a team planning a new mall. Your research has shown that 38% of the mall space should have retail shops, and 16% should have food shops. The rest of the space should be used for walkways, seating areas, indoor gardens or fountains, offices, and restrooms. How will you organize the space for all those things?

Plan
1. Decide on the number of square feet your mall will cover. Base your decision on research you carry out. Start a table with sizes for several mall plans.

2. Figure out how much space in each mall is needed for retail shops and food shops.

Decide As a group, decide on the plan you like best.

Create Make a model or diagram to show your plan for retail and food shops. Indicate where walkways, gardens, fountains, offices, and restrooms will go.

Present As a group, share your plan with the class. Be prepared to explain your thinking.

UNIT 5: Problem Solving Using Geometry and Measurement

Unit Theme: Theme Park

Have you ever noticed all the different shapes that are in and around theme parks? From the gigantic Ferris wheel to the small bricks that line the park's paths, shapes are everywhere. In this unit, you will learn how math is used every day at theme parks.

Math to Know

In this unit, you will apply these math skills:

- Find area, perimeter, and circumference
- Solve problems with similar and congruent figures
- Find surface area and volume of 3-dimensional figures

Problem-Solving Strategies

- Guess, Check, and Revise
- Write an Equation
- Draw a Diagram

Link to the Theme

Write another paragraph about the Wacky Wave Pool. What shape is it? How big is it? How many people can ride in it at one time?

Robin is excited to go to Water World. This year, his cousin Lena is going with him. During the car ride to the park, Robin tells Lena about his favorite ride, the Wacky Wave Pool.

_____
_____
_____
_____
_____

Use Math Language

Review Vocabulary

The list below shows vocabulary terms in this unit. Knowing the meaning of these terms will help you understand the problems.

area	diameter	length	rectangular prism
congruent	edge	Pythagorean theorem	square foot
cubic foot	face	radius	width

Vocabulary Activity   Word Pairs

Math terms are often learned together, but mean different things. Use terms from the above list to complete the following sentences.

A. Solid Figures

1. The flat surfaces of a solid figure are called _____.

2. The line segments where two faces of a solid figure meet are called _____.

B. Circles

3. The distance from the center of a circle to any point on the circle is called the _____.

4. The length of a segment that connects two points on a circle and passes through the center is called the _____.

Graphic Organizer   Word Web

Complete the graphic organizer.

- In each of the three ovals, write a vocabulary word that is related to the word *area*.
- Write a definition for each of the three words.

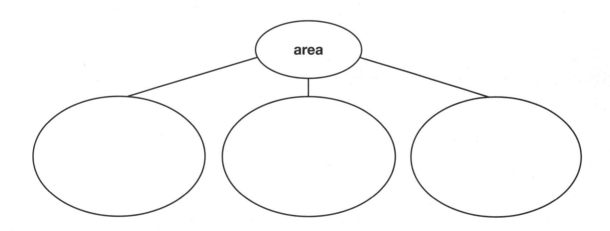

143

Lesson 17

Strategy Focus
Guess, Check, and Revise

MATH FOCUS: Perimeter and Circumference

Learn

Read the Problem

A train runs between the big attractions at a theme park. The theme park is in the shape of a large circle. The track for the train connects points A, B, C, D, E, then back to C and A where it starts the route over again. The whole route is 8,925 feet long. What is the distance from A to E? Use 3.14 for π.

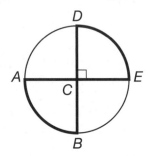

Reread Ask yourself questions as you reread the problem.

- What is the problem about?
  _____

- What do I know?
  _____

- What am I asked to find?
  _____

Mark the Text

Search for Information

Read the problem again. Cross out any details you do not need.

Record Summarize the facts you need.

The theme park is in the shape of a large _____ .

The entire route is _____ feet long.

Think about how you can use this information to solve the problem.

144 Unit 5 **Using Geometry and Measurement**

Decide What to Do

You know that the shape of the route makes a path along parts of a circle. You also know the total length of the route.

Ask How can I find the distance from A to E?

- I can use the strategy *Guess, Check, and Revise* and organize my guesses in a table.
- First, I can guess the length of diameter AE and find the length of the route. I can revise my guesses until I find a length that matches 8,925 feet.

Use Your Ideas

Step 1 Make a table with columns for the parts of the route. Diameters AE and BD are equal. Together, arc AB and arc DE make up half the circumference of the circle. $C = \pi d$, so the sum of their lengths is $\frac{1}{2}\pi d$.

Step 2 Guess a length for the diameter. Then use that length for d in the formulas and complete the other columns of the table.

Step 3 Compare the result to 8,925 to see if the numbers match. If needed, make a new guess and repeat Steps 2 and 3.

Diameter (ft)	Diameter × 2 (ft)	$\frac{1}{2}C = \frac{1}{2}\pi d$ (ft)	Length of Route (ft)	Compare
2,000	4,000	3,140	7,140	too low
3,000	6,000	4,710	10,710	too high
2,500				

The first guess was too low, so try a greater length for the diameter for the next guess.

The distance from A to E is _____ feet.

Review Your Work

Make sure you used the correct numbers in the formulas.

Explain How did knowing the parts of a circle help you?

_____

_____

145

Try

Solve the problem.

① A worker is going to string dinosaur lights around the diner at the theme park. She knows that the rectangular diner has an area of 324 square feet and that the width of the diner is $\frac{1}{4}$ of its length. What are the length and width of the diner?

Mark the Text

▪ Read the Problem and Search for Information

Reread the problem. Draw a diagram of the diner if you need to. Identify the connection between what you know and what you want to find.

▪ Decide What to Do and Use Your Ideas

Use the *Guess, Check, and Revise* strategy. Guess a length and find the width and area. Then you can compare the area to 324 square feet.

Step 1 Make a table to organize your work. Make guesses for the length.

Step 2 Find the width, or $\frac{1}{4}$ of the length. Then find the area. Compare the result to 324 square feet to see if the numbers match.

Ask Yourself
How is the area of the diner related to what I am asked to find?

Length (feet)	$\frac{1}{4}$ of Length (feet)	Area (square feet)	Compare
40	10	400	too high
32			too low

The length of the diner is _____ feet. The width is _____ feet.

▪ Review Your Work

Check that the width you found is $\frac{1}{4}$ of the length.

Describe Why was it helpful to guess a multiple of 4 to start?

_____

_____

Apply

Solve the problems.

(2) A circular pond has bridges that intersect at the center. A circular path surrounds the pond. Mandy walked from point X to point Y and then to point Z. From there she walked back to point X along the path. She walked a total of 714 feet. What is the distance from point X to point Y? Use 3.14 for π.

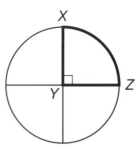

Ask Yourself: What part of the circumference is the length of arc XZ?

Hint $C = \pi d$

Radius (ft)	Radius × 2 (ft)	$\frac{1}{4}$ × Circumference (ft)	Total (ft)	Compare
300	600	471	1,071	

Answer _____

Determine Why is 300 a good first guess for the radius of the pond?

_____

_____

(3) A park has 36 sections of fence to go around a rectangular bumper car floor. Each section of fence is 6 feet long. The area of the floor must be between 2,800 and 2,900 square feet. How many sections should be used on each side of the floor?

Hint The sum of the number of sections of fence on all four sides will be 36.

Width (w) (sections)	Length (l) (sections)	Area (square feet)	Compare
9	9		

Ask Yourself: If I guess the number of sections on one side, how can I find the area?

Answer _____

Apply How does making a table help you to solve the problem?

_____

_____

Lesson 17 **Strategy Focus: Guess, Check, and Revise**

④ A new ride at the theme park is on a triangular plot of land. The area of this right triangle is 651 square meters. The long leg of the triangle is 1 meter shorter than 3 times the length of the short leg. What are the lengths of the legs of the triangular plot of land?

Hint Guess the length of the short leg and draw a diagram.

Ask Yourself
What must I include in a diagram of this plot of land?

Short Leg (m)	Long Leg (m)	Area (m²)	Compare
20			

Answer _____

Relate How did the result of the first guess help you?

_____

_____

⑤ A Ferris wheel has passenger cars attached at the ends of each of 8 diameters. It has a total of 3,899 LED lights. The lights are evenly spaced on the diameters and the circumference. How many lights are on each diameter? Use 3.14 for π.

One diameter

Hint Use the formula for the circumference of a circle.

Ask Yourself
If I guess the number of lights on one diameter, how can I find the number of lights on the whole Ferris wheel?

Answer _____

Sequence Could you solve this problem by first guessing the number of lights on the circumference? Explain.

_____

_____

148   Unit 5   **Using Geometry and Measurement**

Practice

Solve the problems. Show your work.

6 The length of the building that houses a gift shop is 5 times its width. The area of the building is 3,125 square feet. What are the dimensions of the building?

Answer _____

Assess For what kinds of problems is the strategy *Guess, Check, and Revise* most useful?

_____

_____

7 A roller coaster passes through a rectangular gate that opens at the bottom of a steep hill. The perimeter of the gate is 50 feet and the height is 1 foot more than twice its width. What are the dimensions of the gate?

Answer _____

Conclude What is the importance of checking your answer before making another guess?

_____

_____

Create Write and solve a problem that asks about the length and width of a rectangular plot of land. Include the area in your problem. Provide clues about the length and width and check that you can solve the problem using the strategy *Guess, Check, and Revise*.

Lesson 17 **Strategy Focus: Guess, Check, and Revise** 149

Lesson 18

Strategy Focus
Write an Equation

MATH FOCUS: Area and Surface Area

Learn

▢ Read the Problem

> In the fall, a local park will paint the walls and bottom of its pool. The pool has the shape of a rectangular prism. The pool is 50 feet long, 18 feet wide, and 6 feet deep. The painter will put two coats of paint on the pool. How many cans of paint are needed to paint this pool, if one can covers 100 square feet?

Reread Ask questions to be sure you understand the problem.

- What is the problem about?
  _____

- What kind of information is given?
  _____

- What does the problem ask me to find?
  _____

Mark the Text →

▢ Search for Information

Read the problem again and look for the information you need.

Record What are the details that will help you solve the problem?

The pool is _____ feet long, _____ feet wide, and _____ feet deep.

One can of paint covers _____ square feet.

There are to be _____ of paint on the pool.

You can use this information to help you determine how to solve the problem.

150 Unit 5 **Using Geometry and Measurement**

Decide What to Do

You know the dimensions of the pool and how much surface area one can of paint will cover. You also know that the pool has the shape of a rectangular prism and that the paint will cover some of its faces.

Ask How can I find the number of cans of paint needed?

- I can use the strategy *Write an Equation* to find the surface area of the pool. The pool needs 2 coats of paint, so I will multiply the surface area by 2 to find the total surface area to be painted.

- Then I will divide by the area each can of paint covers to find the total number of cans needed.

Use Your Ideas

Step 1 Write an equation to find the surface area. The pool has 5 faces: the bottom, 2 long walls, and 2 short walls.

Surface Area = Area of Bottom + 2 Long Walls + 2 Short Walls
$SA = (50 \times 18) + 2(50 \times 6) + 2(18 \times 6)$
$ = 900 + 600 + 216$
$SA = 1{,}716$ square feet

> You can use formulas you know to help you solve the problem.

Step 2 Multiply the surface area by 2.

$2 \times SA = 2 \times 1{,}716$

= _____

For 2 coats, a total of _____ square feet must be painted.

Step 3 Divide 3,432 by 100. _____

Look at your quotient. Does it make sense to include a fraction of a can of paint in your answer?

So a total of _____ cans of paint will be needed.

Review Your Work

Check that you have answered the question that was asked.

Describe How does knowing the shape of the pool help you?

_____

_____

Try

Solve the problem.

1 A sculpture in a garden is shaped like a cube that is balanced on one corner. A group of students is making a full-size model of the cube. They use exactly 54 square feet of cardboard to make the outside of the cube. Each edge of the cube will be decorated with a ribbon. How long should each ribbon be?

Read the Problem and Search for Information

Identify the details given in the problem and decide if there is information you need that is not in the problem. Visualize the problem to be sure you know what question is being asked.

Decide What to Do and Use Your Ideas

You can use the strategy *Write an Equation* to solve the problem.

Step 1 Use the formula for the surface area of a cube. Let s represent the length of one edge.

$$\text{Surface Area} = 6 \times s^2$$
$$54 = 6 \times s^2$$

Ask Yourself
What is the formula for the surface area of a cube?

Step 2 Solve the equation.
$$\text{Surface Area} = 6 \times s^2$$
$$54 = 6 \times s^2$$
$$\frac{54}{6} = \frac{6 \times s^2}{6} \quad \leftarrow \text{Divide both sides by 6.}$$
$$\underline{\qquad} = s^2$$

To solve for s, find what number multiplied by itself equals 9.

$s = $ _____

Each piece of ribbon should be _____ long.

Review Your Work

Does your answer work when you substitute it into the formula?

Explain A friend says that every formula is an equation. Do you agree? Why or why not?

_____

_____

Apply

Solve the problems.

2 The path to a castle in a theme park needs to be paved with yellow bricks. It takes 8 bricks to pave each square foot. How many bricks will it take to pave the path with the dimensions shown in the diagram?

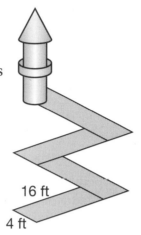

Hint The path is made of 4 large congruent rectangles.

The area of each rectangle in the path is _____ square feet.

The total area of the path is _____ square feet.

Ask Yourself
What operation(s) do I need to use to solve the problem?

Answer _____

Employ Could you write a single equation that solves the problem? What would be the advantage?

_____

_____

3 A local park has a square pyramid sculpture that sits on the ground. Workers are painting the sides. There are 192 square feet to be painted. The workers need ladders tall enough to reach to the top of each face of the pyramid. The sides of the square base are 12 feet long. What is the height of each triangular face?

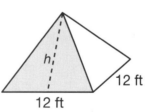

Ask Yourself
Will there be any paint used on the base?

The surface area is _____ square feet.

There are _____ triangular faces to be painted.

The area of each triangular face is _____ square feet.

Hint The formula for the surface area of a square pyramid is $SA = 4(\text{Area of 1 triangular face}) + \text{Area of the base}$.

Answer _____

Apply Why is it sometimes necessary to modify a formula before you use it?

_____

_____

Lesson 18 Strategy Focus: Write an Equation   153

Hint Find the area of the entire picnic area.

4 A rectangular picnic area is being made at a park. The picnic area will include the features shown below. The rest of the picnic area will be planted with grass. Each bag of grass seed covers 210 square feet. If the outside dimensions of the picnic area are 250 feet by 180 feet, about how many bags of grass seed will be needed? Use 3.14 for π.

Feature	Size
Circular fountain	Diameter = 20 feet
Wooded area	60 feet by 40 feet

Ask Yourself

How many square feet will not be covered by grass?

Area of fountain ≈ _____ square feet

Wooded area = _____ square feet

Answer _____

Analyze How do you know which operations to use in the problem?

_____

_____

Hint The area of the pool is equal to the area of one whole circle plus the area of a rectangle.

5 A park has a pool with colorful fish. The dimensions of the pool are shown. A deck will surround the pool so that people can stand and look at all the fish. The deck is shown by the shaded area. The outside edge of the deck will form a rectangle. The inside edge will fit tightly around the edge of the pool. What will the area of the deck be? Use 3.14 for π.

Ask Yourself

How can I find the area of the deck after I know the area of the pool?

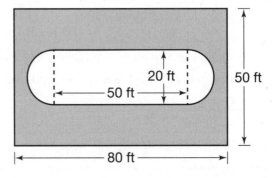

Area of the pool ≈ _____ square feet

Answer _____

Compare Can you begin by finding the area of the deck? Explain.

_____

_____

154 Unit 5 **Using Geometry and Measurement**

Practice

Solve the problems. Show your work.

6 A local group will paint the outside walls of a museum. The building is shaped like a rectangular prism with an overall length of 100 feet and a width of 75 feet. The total area of the windows and doors is 2,400 square feet. The building is 20 feet high. A can of paint covers about 300 square feet. How many cans of paint are needed to put one coat of paint on the outside of the museum?

Answer _____

Consider How is this problem like the first problem in the lesson? What is one way in which the problems are different?

_____

_____

7 A theme park has a site where visitors can dig for artificial dinosaur bones. The site is in the shape of a trapezoid, as shown. A wall will be built around the perimeter of the site. If the surface area of the inside of the wall is 352 square feet, what is the height of the wall?

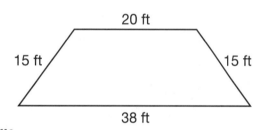

Answer _____

Decide Which dimension is the unknown in your equation?

_____

_____

Create Write a problem about painting a room in a museum that can be solved by using the strategy *Write an Equation*. Provide the dimensions of the room. Include the number of square feet that can be covered with one can of paint. Solve your problem.

Lesson 18 **Strategy Focus: Write an Equation** 155

Lesson 19

Strategy Focus: Draw a Diagram

MATH FOCUS: Similar and Congruent Figures

Learn

Read the Problem

The Visitor Center at Ride City is making a map of the park. The park is in the shape of an isosceles trapezoid with bases of length 10,000 feet and 16,000 feet and sides of length 5,000 feet. On the map, the longer base of the trapezoid will be 64 inches long. What should the height of the trapezoid be on the map?

Reread Check that you have correctly understood the problem.

- What is the problem about?

  _____

- What details do you know?

  _____

- What are you asked to find?

  _____

Mark the Text

Search for Information

Find the details that you need in the problem.

Record Write the information that you found.

The lengths of the bases of the actual park are _____ and

_____ .

The lengths of the sides of the actual park are _____ and

_____ .

The length of the longer base of the trapezoid on the map is

_____ .

Use this information to help you decide how to solve the problem.

156   Unit 5 **Using Geometry and Measurement**

Decide What to Do

You know the shape and dimensions of the actual park. You also know one dimension of the shape on the map of the park.

Ask How can I find the height of the trapezoid on the map?

- I can use the strategy *Draw a Diagram* to show the park.
- Next, I can use the Pythagorean theorem to find the height of the actual trapezoid. Then I can use a proportion to find the height of the trapezoid on the map.

Use Your Ideas

Draw a diagram of the park and label the dimensions you know.

Step 1 Use the Pythagorean theorem to find the height of one of the congruent right triangles. Label the height on the diagram.

$a^2 + b^2 = c^2$

$3{,}000^2 + h^2 = 5{,}000^2$

$9{,}000{,}000 + h^2 = 25{,}000{,}000$

$\qquad h^2 = 25{,}000{,}000 - 9{,}000{,}000$

$\qquad h^2 = 16{,}000{,}000$

$\qquad h = \underline{\qquad\qquad}$

> The trapezoid is made up of a rectangle and two congruent right triangles, so you know the base of each triangle must be 3,000 feet.

Step 2 Use what you know to write and solve a proportion.

$$\frac{\text{Height on map}}{\text{Height of actual trapezoid}} = \frac{\text{Base on map}}{\text{Base of actual trapezoid}}$$

$$\frac{a}{4{,}000} = \frac{64}{16{,}000}$$

$$16{,}000a = 256{,}000$$

$$a = 16$$

The height of the trapezoid on the map should be _____ inches.

Review Your Work

Check that your answer is reasonable.

Relate How did writing the dimensions on the diagram help you?

_____

_____

Try

Solve the problem.

1. A 32-foot by 8-foot garden is divided into 6 smaller gardens by paths. The small gardens are similar to the large garden and congruent to each other. If the length of a small garden is 8 feet, what is its width? How wide is the path?

■ Read the Problem and Search for Information

Visualize the garden. Identify the dimensions given in the problem.

■ Decide What to Do and Use Your Ideas

Draw a diagram and label the dimensions you know.

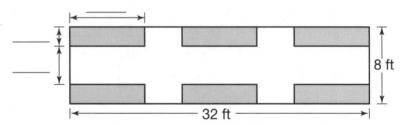

Step 1 Write and solve a proportion to find the width of each small garden. Label the width on the diagram.

$$\frac{\text{Width of small garden}}{\text{Width of large garden}} = \frac{\text{Length of small garden}}{\text{Length of large garden}} \rightarrow \frac{w}{8} = \frac{8}{32}$$

$$32w = 64$$

$$w = \underline{}$$

The width of each small garden is _____.

Step 2 Find the width of the path. Label it on the diagram.
Width of large garden − 2(Width of small garden) = _____

The width of each small garden is _____ and the width of the path is _____.

Ask Yourself

How can I find how much room there is for the path between the small gardens?

■ Review Your Work

Check that your answers work with the dimensions given in the problem.

Conclude How did adding to the diagram as you found new information help you solve the problem?

_____

_____

158 Unit 5 **Using Geometry and Measurement**

Apply

Solve the problems.

2 The scale model for a roller coaster has a section of track that is 18 inches long with a drop of 8 inches. The actual drop of that section of the roller coaster is 160 feet. How long is the actual track for that section?

> **Hint** Draw a diagram for the roller coaster.

Model **Roller Coaster**

> **Ask Yourself**
> How are the diagrams of the model and the roller coaster related?

$\dfrac{8 \text{ inches}}{160 \text{ feet}} = $ _____

Answer _____

Apply How can using a diagram help you write a proportion?

_____

_____

3 A park is shaped like the right triangle below. Inside the park is a triangular pool. The shape of the pool is similar to that of the park. If the shortest side of the pool is 6 feet long, what is the area of the pool?

> **Hint** Draw the pool inside the diagram of the park.

Let x stand for the length of the side that corresponds to 40 feet.

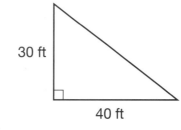

> **Ask Yourself**
> How can I find the length of the other leg of the similar right triangle?

$\dfrac{6 \text{ feet}}{30 \text{ feet}} = $ _____

Answer _____

Interpret How can you use your diagram to check that your answer is reasonable?

_____

_____

Lesson 19 **Strategy Focus: Draw a Diagram**

Ask Yourself

What is the diameter of the circular fence around the *outside* of the actual moat?

Hint The diameter of the inner circle in the diagram is 190 feet.

④ A moat surrounds a castle in the park. The moat is 10 feet wide and 4 feet deep. A fence around the moat prevents accidental falls into the moat. The kit for the scale model that is sold in the park store uses a scale of 1 inch = 10 feet. On the scale model, how many inches of fence will represent the length of the actual fence around the moat? Use $\frac{22}{7}$ for π.

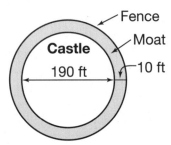

Circumference of actual fence ≈ _____

Answer _____

Examine How does using a diagram help you find dimensions that are not given in a problem?

_____

_____

Ask Yourself

How can I find the lengths of the parts of the route that are not labeled on the figure?

Hint The diagonals of a rhombus bisect each other.

⑤ A park has the shape of rhombus ABCD. A parade leaves from point A, marches to point B, then through the plaza to point D, then to point C, and back through the plaza to point A. The roads through the center of the plaza form four congruent triangles. A scale model has the dimensions shown in the figure. The actual road from point A to point B is 400 feet long. How long is the entire actual parade route?

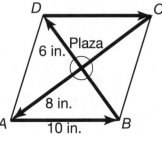

_____ = _____

Answer _____

Explain How did you find the length of the parade route on the model?

_____

_____

Practice

Solve the problems. Show your work.

6 A popular attraction at a theme park is the House of Illusions. Two walls in a room inside the house are parallelograms, each having the dimensions shown below. In the center of each of these two walls is a mirror that is similar to the shape of the wall. The base of each mirror is 6 feet long. What is the perimeter of each mirror?

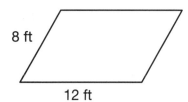

Answer _____

Analyze How would you explain to someone the advantages of using a diagram to solve a problem?

_____

_____

7 A park is shaped like a rectangle. It is 750 feet long and 550 feet wide. A map of the park uses a scale of 1 inch = 50 feet. What is the area of the park on the map?

Answer _____

Identify Explain how you used the scale to solve the problem.

_____

_____

Create

Write a problem about a ride at a theme park that can be solved using the *Draw a Diagram* strategy. Solve your problem.

Lesson 19 **Strategy Focus: Draw a Diagram** 161

Lesson 20

Strategy Focus
Write an Equation

MATH FOCUS: Volume and Capacity

Learn

Read the Problem

A water park has a circular children's pool with a diameter of 20 feet. A worker must fill the pool to a depth of 2 feet. Water can be added at the rate of 157 cubic feet per hour. If the park opens at 10:00 A.M., what time does the worker need to start filling the pool so it will be full by the opening time? Use 3.14 for π.

Reread Ask yourself questions as you read the problem.

- What is the problem about?
  _____

- What measurements are given?
  _____

- What other kind of information do you know?
  _____

- What are you asked to find?
  _____

Mark the Text

Search for Information

Read the problem again. Find the facts that you need to solve the problem.

Record What details do you need to answer the question? Write the facts you know.

The diameter of the pool is _____ .

The pool will be filled with water to a depth of _____ .

Water can be added at the rate of _____ .

The water park opens at _____ .

Think about how you can use this information to help you choose a strategy.

162 Unit 5 Using Geometry and Measurement

Decide What to Do

You know the diameter of the pool and its depth. You also know how fast water can be added to the pool.

Ask How can I find when the worker needs to start filling the pool?

- I can use the strategy *Write an Equation* to find the volume of the pool.
- Next, I can divide the volume by 157 cubic feet per hour to find how many hours it takes to fill the pool. Then I can find what time the worker needs to start.

Use Your Ideas

Step 1 Write and solve an equation to find the volume of the pool.

$V = \pi r^2 h$

$\approx 3.14 \times 10^2 \times 2$

$\approx \underline{\qquad}$

The volume of the pool is about _____ cubic feet.

Step 2 Divide the volume of the pool by 157 to find about how long it will take to fill the pool.

_____ cubic feet ÷ 157 cubic feet per hour = _____ hours

It will take about _____ hours to fill the pool to a depth of 2 feet.

Step 3 Find the time the worker needs to start filling the pool.

_____ hours before 10:00 A.M. is _____ .

So the worker must start filling the pool at _____ .

> The pool is the shape of a cylinder. You can use the formula $V = \pi r^2 h$ to find the volume of a cylinder.

Review Your Work

Make sure you substituted the correct values in the equation.

Explain How does the strategy *Write an Equation* help you?

_____

_____

Try

Solve the problem.

1 A restaurant uses about 1,200 pounds of ice each day during the summer months. One cubic foot of water makes about 60 pounds of ice. About how many cubic feet of water are needed each day to make ice for the restaurant?

▢ Read the Problem and Search for Information

Think about the question you need to answer. Then ask yourself if there is other information you will need.

▢ Decide What to Do and Use Your Ideas

You know how many pounds one cubic foot of ice makes and how many pounds of ice are needed. You can use the strategy *Write an Equation* to find how many cubic feet of water are needed.

Ask Yourself: What two ratios should I write in my proportion?

Step 1 Write a proportion. One cubic foot of water makes about _____ pounds of ice. Let x stand for the number of cubic feet of water needed to make 1,200 pounds of ice.

$$\frac{1}{60} = \underline{\qquad} \quad \begin{array}{l}\leftarrow \text{cubic feet of water} \\ \leftarrow \text{pounds of ice}\end{array}$$

Step 2 Solve the proportion.

$$\frac{1}{60} = \frac{x}{1{,}200}$$

$$\underline{\qquad} = \underline{\qquad}$$

$$\underline{\qquad} = \underline{\qquad}$$

The restaurant will need about _____ cubic feet of water each day.

▢ Review Your Work

Substitute your value of x into the proportion. Are the ratios equal?

Recognize Suppose the restaurant uses about 600 pounds of ice each day during the summer. Can you still use the same ratio in your proportion to find about how many cubic feet of water are needed to make ice? Explain why or why not.

_____

_____

Apply

Solve the problems.

(2) A cylindrical water tower in Town A has a capacity of about 6,280 cubic feet of water. Town B has a water tower that holds about 46,900 gallons of water. There are about 75 gallons in 10 cubic feet of water. Which water tower has the greater capacity?

Ask Yourself
Will I compare gallons of water or cubic feet of water?

$\dfrac{10}{75} = \dfrac{}{}$ ← cubic feet of water
← gallons of water

$x = $ _____

Hint Use x to stand for the number of gallons of water in the tower in Town A.

Answer _____

Distinguish Why do you use a proportion instead of a formula for the volume of a cylinder to find how many gallons of water the tower in Town A holds?

_____

_____

(3) A pool in the water park is in the shape of a rectangular prism. It is 20 feet long and 15 feet wide. There are about 1,800 cubic feet of water in the pool. Joan's parents do not want her to jump in a pool where the water is less than 3 feet deep. Is the water deep enough for Joan to jump in? How many feet deep is it?

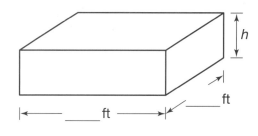

Hint Label the dimensions that you know on the diagram.

$1{,}800 = $ _____ × _____ × _____

Ask Yourself
What must I solve for in my equation?

Answer _____

Describe Which equation did you use to solve the problem? Why did you choose this equation?

_____

_____

Lesson 20 **Strategy Focus: Write an Equation** 165

Ask Yourself

How can I use the volume to find the number of Btu needed each hour?

Hint Let x stand for the number of Btu needed to air condition the store.

④ A store is the shape of a rectangular prism. It is 120 feet long, 80 feet wide, and 12 feet high. Suppose a worker at the store finds that a volume of 100,000 cubic feet uses 300,000 British thermal units (Btu) of air conditioning per hour. Based on that, how many Btu per hour would be needed to air condition the store?

Volume of the store: _____

▶ $\dfrac{100{,}000}{300{,}000} = \underline{}$

Answer _____

Sequence What steps did you take to solve the problem?

_____

_____

Ask Yourself

If I know the number of minutes, how can I find the equivalent number of hours?

Hint The pool is a rectangular prism plus one cylinder. Use 3.14 for the value of π.

⑤ A swimming pool is in the shape of a rectangular prism with half cylinders of the same size at each end. Workers will pump water into the pool at a rate of 15 cubic feet per minute. About how long will it take to fill the pool to a depth of 5 feet? Can the workers fill the pool in one 8-hour work day? Explain.

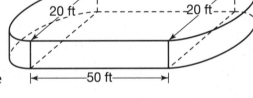

▶ Volume of rectangular prism: _____

Volume of cylinder: about _____

Answer _____

Explain The problem says the pool has ends shaped like half cylinders. Why can you use the formula for the volume of a cylinder instead of modifying it?

_____

_____

166   Unit 5   **Using Geometry and Measurement**

Practice

Solve the problems. Show your work.

6 A park has a pool in the shape of a rectangular prism. The pool is 20 feet long and 12 feet wide. A lifeguard chair is built into one end of the pool. The distance from the bottom of the pool to the first step on the chair is $1\frac{3}{4}$ feet. If the pool has 360 cubic feet of water, can a lifeguard stand on the first step of the chair without being in the water? Why or why not?

Answer _____

Adapt A pool has the shape of a rectangular prism. Suppose you know the length of the pool, the depth of the water, and the pool's volume, but not its width. How can you use the equation for the volume of a rectangular prism to find the width?

_____

_____

7 A restaurant has a walk-in refrigerator in the shape of a rectangular prism. It has an interior length of 12 feet and width of 10 feet. The volume of the refrigerator is 720 cubic feet. A chef at the restaurant is 6 feet 2 inches tall. What is the difference between her height and the interior height of the refrigerator? Can she walk in without bending?

Answer _____

Infer Why does the problem give the interior length and width of the refrigerator instead of its outside length and width?

_____

_____

Create Write a problem about a swimming pool in the shape of a cylinder. Give the height and diameter. Ask about the volume of the pool. Then solve your problem using the strategy *Write an Equation*.

Unit 5 Review

In this unit, you worked with three problem-solving strategies. You can often use more than one strategy to solve a problem. So if a strategy does not seem to be working, try a different one.

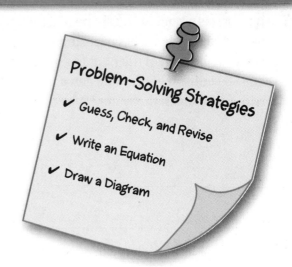

Problem-Solving Strategies
- ✓ Guess, Check, and Revise
- ✓ Write an Equation
- ✓ Draw a Diagram

Solve each problem. Show your work. Record the strategy you use.

1. People gather at a circular fountain in the park to watch a light show on the water. They can sit on the 314-foot long wall that surrounds the fountain. How far from the center of the fountain are the people sitting on the wall? Use 3.14 for π.

2. A rectangular garden with an area of 64 square feet is 4 times as long as it is wide. What is the perimeter of the garden?

Answer _____

Strategy _____

Answer _____

Strategy _____

3. An artist is making an ice sculpture from a rectangular block of ice with a base of 2 feet by 3 feet. A cubic foot of ice weighs about 60 pounds. If the artist's block of ice weighs 1,440 pounds, how tall is it?

Answer _____

Strategy _____

4. A restaurant has tables that can seat four people on each side and one person on each end. A customer wants the tables arranged end to end with guests seated only on the outer perimeter. How many tables are needed to seat 42 people?

Answer _____

Strategy _____

5. A scale model for a proposed shopping mall uses a scale of 1 inch = 10 feet. What is the area of the rectangular floor of a store that is 12 inches by 10 inches on the scale model?

Answer _____

Strategy _____

Explain how you used the strategy to solve this problem.

Solve each problem. Show your work. Record the strategy you use.

6. A rectangular dog run in a park is enclosed by 32 feet of fence. The area of the dog run is 60 square feet. What are its dimensions?

Answer _____

Strategy _____

7. A courtyard between apartment buildings is in the shape of a trapezoid with 2 right angles. The parallel bases are 1,400 and 1,700 feet long. The length of the slanted side connecting the bases is 500 feet. What is the area of the courtyard?

Answer _____

Strategy _____

8. A 628-square foot mural is being designed to go completely around the side of a cylindrical water tower. The radius of the tower is 10 feet. What is the height of the mural? Use 3.14 for π.

Answer _____

Strategy _____

Explain how you had to modify the formula for the surface area of a cylinder to solve the problem.

170   Unit 5   Using Geometry and Measurement

9. A walk-in freezer at a grocery store has the shape of a rectangular prism. The freezer has a volume of 960 cubic feet. The interior width is 5 feet and the height is 8 feet. What is the length of the interior?

Answer _____

Strategy _____

10. A rectangular deck has dimensions of 8 feet by 12 feet. One of the longer sides borders a building. A 2-foot wide garden surrounds the other three sides. What is the area of the garden?

Answer _____

Strategy _____

Write About It

Look back at Problem 10. Describe how you used the information in the problem to choose a strategy for solving the problem.

_____

_____

_____

Team Project: Design a Park

Your team has been asked to design a community park. The park will be 500 feet wide and 1,000 feet long. Create spaces for at least three different features, from the list at the right, each with a different shape. Make a scale drawing of your park design and label the dimensions.

Possible Park Features
Ballpark
Picnic Area
Outdoor Theater
Tennis Courts
Fountain
Play Area

Plan
1. You must include spaces for at least three features. For each feature, show the dimensions and the area of the space.
2. The scale drawing must include a key to show the scale.

Decide As a group, choose the features and the scale you will use.

Create Make a scale drawing of the park showing the position and size of the space for each feature.

Present As a group, share your design with the class.

UNIT 6: Problem Solving Using Data and Graphing

Unit Theme:
The Great Outdoors

There are so many activities to enjoy when you are outdoors. You might go on a camping trip or for a swim in a pool. Or maybe you like listening to a concert in the park. Whether you want to relax or be active, there is always something to do. In this unit you will see how math is part of the great outdoors.

Math to Know

In this unit, you will apply these math skills:

- Interpret data presented in various forms
- Draw conclusions from double bar graphs and double line graphs
- Make and use circle graphs, box plots, and stem-and-leaf plots

Problem-Solving Strategies

- Use Logical Reasoning
- Make a Graph
- Make an Organized List

Link to the Theme

Write another paragraph about the skateboard classes. Include some of the words and numbers from the table.

Lani wants to learn how to skateboard. She finds information about classes her city offers. She shows the information to her parents.

_____

_____

_____

_____

_____

_____

Class Length	Prices
1 Day	$12
3 Days	$30
5 Days	$50
Skateboard, helmet, and pads included.	

Use Math Language

Review Vocabulary

The list below shows vocabulary terms in this unit. Knowing the meaning of these terms will help you understand the problems.

central angle	double line graph	measure of central tendency	outlier
circle graph	frequency table	median	quartile
double bar graph	mean	mode	scale

Vocabulary Activity   Modifiers

A descriptive word placed in front of another word can indicate a specific meaning. When learning new math vocabulary, pay attention to each word in the term.

1. I can look at the length or height of the bars in a _____ to compare two sets of data.

2. When I have to show how two sets of data change over time, I can use a _____ .

3. To compare parts of a whole, I can use a _____ .

4. To show how often something occurs, I can organize data in a _____ .

Graphic Organizer   Word Circle

Complete the graphic organizer.

- Cross out the word that does not belong.
- Replace it with a word from the vocabulary list that does belong.
- Write a definition for each word.

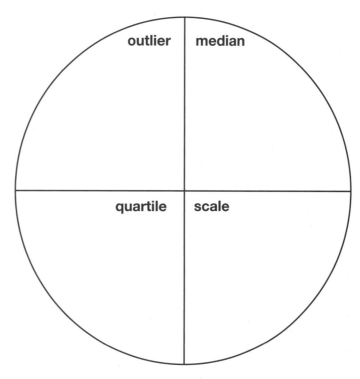

Lesson 21

Strategy Focus
Use Logical Reasoning

MATH FOCUS: Interpret Data

Learn

▣ Read the Problem

Ms. Lopez works at the gift shop at a national historic park. She made this stem-and-leaf plot to show the prices of the items at the souvenir counter. She thinks two prices are outliers and removes them from the data set. How does removing the outliers affect the mean, median, and mode of the prices?

Souvenir Prices

Stem	Leaf
0	75
1	50 50 75
2	25 50 75
3	25 50 75
4	25 25
5	50
6	
7	
8	
9	50 95

Key: 0 | 75 = $0.75

Reread Ask yourself these questions as you read the problem again.

- What kind of data has Ms. Lopez collected?

  _____

- What question am I asked to answer?

  _____

▣ Search for Information

Read the problem again and mark the information that will help you find a solution.

Record Write what you know about the souvenir prices.

There are _____ items with a price in the stem-and-leaf-plot.

The range of the prices is from _____ to _____ .

List the prices in the table. _____

_____

Think about how you can use this information to help you.

174   Unit 6   **Using Data and Graphing**

Decide What to Do

You know the prices of 15 items at the souvenir counter. You know two of the prices are outliers.

Ask How can I compare the mean, median, and mode of the data set before and after the outliers are removed?

- I can use the strategy *Use Logical Reasoning* and my knowledge about stem-and-leaf plots to find differences in the measures of central tendency.

- I can find the mean, median, and mode of all the prices. Next, I can identify the outliers and find the mean, median, and mode of the rest of the prices.

- Then I can compare the two sets of mean, median, and mode to find how each changes.

Use Your Ideas

Step 1 Find the mean, median, and mode of all the prices.

Mean = _____   Median = _____

Mode = _____

> Remember there can be more than one mode.

Step 2 Which two prices are the outliers? _____

Step 3 Find the mean, median, and mode of the prices excluding the outliers.

Mean = _____   Median = _____

Mode = _____

So the mean _____ by _____ , the median _____ by _____ , and the mode _____ .

Review Your Work

Check to be sure that you included all the souvenir prices from the stem-and-leaf plot in your calculations.

Describe Refer to your work in Steps 1 and 3. Which measure is most affected by the presence and removal of the outliers, the mean or the median? Explain.

_____

_____

Try

Solve the problem.

Ask Yourself
How are these graphs the same? How do they differ?

1 Mr. Alou works at the National Lakeshore. He plots the data from March and July for the number of tent campers per day. Using his box plots, he is going to report that the median numbers of tent campers in March and July were about the same. What should he report instead?

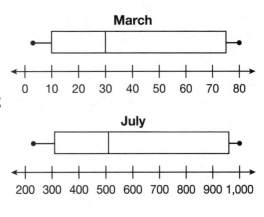

Mark the Text

▪ Read the Problem and Search for Information

Think about what you know about making and using box plots.

▪ Decide What to Do and Use Your Ideas

Use logical reasoning as you compare the graphs.

Step 1 Compare the scales used on the box plots. Are the scales the same? _____

Step 2 Find the median number of campers for both March and July.

The median for the March plot is _____ .

The median for the July plot is about _____ .

So Mr. Alou should report that the median number of campers for March was _____ and for July was _____ .

▪ Review Your Work

Look back at the box plots to be sure you read the data correctly.

Identify Mr. Alou thought the medians, upper quartiles, and lower quartiles of both months were about the same. Why might he think this?

_____

_____

176 Unit 6 **Using Data and Graphing**

Apply

Solve the problems.

2 A national park had to close several times last year. Two workers made box plots to represent the number of days of each closing. Mr. Green thinks they should use his plot because it shows that the median number of days is less. Is he correct? Explain.

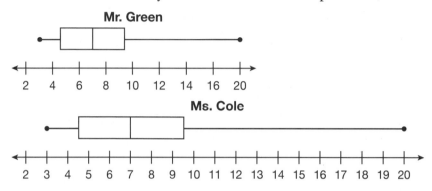

> **Hint** The quartiles are the left and right edges of each box.

	Median	Upper Quartile	Lower Quartile	Range
Mr. Green		9.5	4.5	
Ms. Cole				

Ask Yourself
How will the table help me to compare the box plots?

Answer _____

Explain How did the table help you to solve this problem?

_____

_____

3 Mr. Khan timed his students as they climbed a hill. His data are shown in the stem-and-leaf plot. The first 4 students ran up the hill. How did their times affect the mean, median, and mode?

Times in Seconds

Stem	Leaf
5	8
6	0 0 2
7	
8	5 5 7 8 9
9	7 8 8
10	5 5 7 9 9 9
11	1
12	0

Key: 8 | 5 = 85 seconds

Ask Yourself
Which four times in the stem-and-leaf plot represent the four students who ran?

	All Students	All But the 4 Runners
Mean	92.1 sec	
Median	97.5 sec	
Mode		

> **Hint** The median is the middle number in a set of data when the data are ordered.

Answer _____

Select Which measure needs no calculation? Explain.

_____

_____

Ask Yourself

How can I tell if the graph is accurate?

Hint Look carefully at the axes of the graph.

④ Anton made a bar graph that shows the number of national forests in each state he plans to visit this summer. Paul says that California has 3 times as many National Forests as Montana. Explain how you know that Paul's statement is not correct.

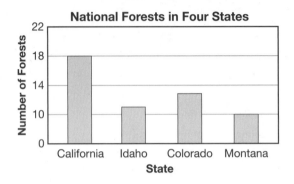

California has _____ national forests.

Montana has _____ national forests.

Answer _____

Demonstrate How could you make the graph more accurate?

_____

_____

Ask Yourself

Which values from the table did Chang-Sun drop?

Hint In this case, the median will be the average of the two middle numbers.

⑤ Chang-Sun found data about visitors to the New Orleans Jazz National Historic Park between 2000 and 2009. He wondered how the mean, median, and mode would change if he removed the least and greatest values. Describe the effect that removing those two values would have on the mean, median, and mode.

Year	2000	2001	2002	2003	2004
Visitors	16,711	45,766	50,019	43,926	44,226

Year	2005	2006	2007	2008	2009
Visitors	40,242	34,375	49,053	66,726	80,828

Answer _____

Compare How is this problem similar to the Learn problem?

_____

_____

178 Unit 6 **Using Data and Graphing**

Practice

Solve the problems. Show your work.

6 Ruth made a bar graph showing information about the number of areas protected by the National Park Service in five western states. Ruth says that Arizona has more than twice as many National Park Service areas as any of the other states. Is Ruth's conclusion correct? Explain your reasoning.

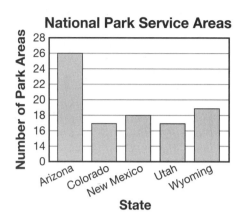

Answer _____

Determine What is a true comparison Ruth can make about the data on this graph?

_____

_____

7 Olivia's class took a trip to the Washington Monument. Each student made a guess about the number of blocks in the monument. Olivia recorded the guesses in a stem-and-leaf plot. If she removes the outlier from the guesses, how will the mean, median, and mode of the data set be affected?

Guesses for Number of Blocks (thousands)

Stem	Leaf
3	7 8
4	0 5 8
5	
6	2
7	0 0 5
8	
9	4

Key: 9 | 4 = 94,000

Answer _____

Formulate What is another question this problem could ask?

_____

_____

Create

Write a problem that uses a stem-and-leaf or a box plot and that can be solved by using logical reasoning. Solve your problem.

Lesson 21 **Strategy Focus: Use Logical Reasoning** 179

Lesson 22

Strategy Focus
Make a Graph

MATH FOCUS: Circle Graphs and Bar Graphs

Learn

Read the Problem

Natalie's family enjoys bird watching. Each December, they keep track of the number of birds they see during the month. Natalie wonders how the number of birds they saw in 2009 compares to the number of birds they saw 2008. Which bird sighting increased the most from 2008 to 2009?

2008	
Hermit Thrush	55
Black-capped Chickadee	80
Song Sparrow	45
Purple Finch	70
Northern Flicker	30

2009	
Song Sparrow	50
Northern Flicker	20
Black-capped Chickadee	70
Purple Finch	80
Hermit Thrush	40

Reread Ask yourself these questions as you read.

- What is Natalie wondering about?

  _____

- What question do you need to answer?

  _____

Mark the Text

Search for Information

Read the problem again. Circle any words that will help you decide how to solve the problem.

Record Write what you know about the data that Natalie's family collected.

How many types of birds does Natalie's family count each year? _____

How many years of data does Natalie want to compare? _____

What is the least number of birds they saw in either year? _____

What is the greatest number of birds they saw in either year? _____

Use the data to help you decide how to solve the problem.

180 Unit 6 **Using Data and Graphing**

Decide What to Do

You know the types of birds Natalie's family counts each year and the numbers of each type they see. You have data for two years.

Ask How can I find out which bird sighting increased the most?

- I can use the strategy *Make a Graph* to organize the data. I can make a double bar graph to display the two sets of data.

- I can compare the bars for each type of bird to decide which bird sightings increased the most from 2008 to 2009.

Double bar graphs are useful for comparing two sets of data.

Use Your Ideas

Step 1 Draw and label the axes for your graph.

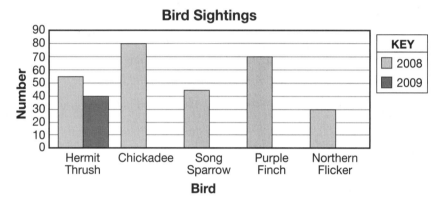

Step 2 Draw the bars. Use a different color for each year and make a key. Complete the graph.

Step 3 Use the graph to see which bird had the greatest increase.

Sightings of chickadees and northern flickers each decreased by _____ .

Sightings of hermit thrushes decreased by _____ .

Sightings of song sparrows increased by _____ .

Sightings of purple finches increased by _____ .

So the bird sighting that increased the most was the _____ .

Review Your Work

Check your graph to make sure the heights of the bars match the data.

Describe How does making a double bar graph help you?

_____

_____

Try

Solve the problem.

1. The circle graph shows the results for a group of campers who ran a 1-mile race at the beginning of the summer. The table shows the results for the same group of campers at the end of the summer. In general, do the times improve or worsen?

Beginning of Summer

End of Summer
Under 35 minutes: 5%
35–40 minutes: 75%
41–45 minutes: 15%
Over 45 minutes: 5%

Read the Problem and Search for Information

Reread the problem. Study the graph and the data.

Decide What to Do and Use Your Ideas

Ask Yourself

How do I find the measures of the central angles to use for each section?

You can make a circle graph to show the end of summer data.

Step 1 Find the measures of the central angles.

Under 35 minutes:
5% × 360° = _____

35–40 minutes:
75% × 360° = _____

41–45 minutes:
15% × 360° = _____

Over 45 minutes: 5% × 360° = _____

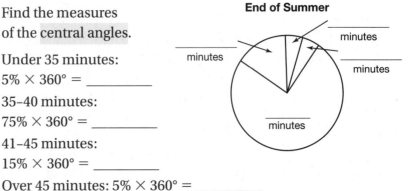

End of Summer

Step 2 Complete the circle graph. Make comparisons between the two circle graphs.

In general, the times _____ .

Review Your Work

Be sure that the sum of the measures of the central angles is 360°.

Summarize How can a double bar graph help solve this problem?

_____

_____

Apply

Solve the problems.

2 A 5K race is about 3.1 miles long. The table shows the times and ages of people who finished in under 30 minutes. The data is reported in the age ranges shown in the graph. Which age group had the most people finish in under 30 minutes?

Time	Age
15:46	28
15:47	24
15:59	43
16:40	15
18:55	44
20:02	17
20:12	23
20:30	55
22:50	58
28:10	33
29:33	44
29:42	19

Hint A time of 15:46 means that the runner finished in 15 minutes and 46 seconds.

Ask Yourself What kind of graph can I make from the data?

Under-30-Minute Finishers (bar graph with Age categories: 15 and Under, 16–21, 22–30, 31–44, 45 and over)

Answer _____

Identify What information is not needed to solve the problem?

_____

_____

3 Two hundred fish were caught. Half were less than 8 inches long. 25% were between 8 and 10 inches long. 20% were between 10 and 12 inches long. 5% were 12 inches or longer. Which size fish was caught most often? How many fish of that size were caught?

Ask Yourself If half of the fish were less than 8 inches long, what percent is that?

Hint Use the percents to find the angle measures.

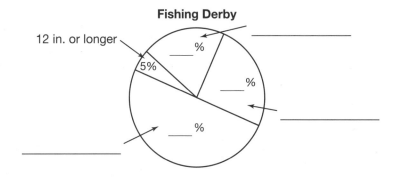

Fishing Derby

Answer _____

Generalize How did the graph help you?

_____

_____

Lesson 22 **Strategy Focus: Make a Graph** 183

Hint Find the total number of runners in last year's run.

④ Mr. Steele made a circle graph showing the runners in different age groups in last year's 5K run. There are 250 runners this year. If the age groups in both years have the same percents of runners, how many runners this year are in the 19–24 age group?

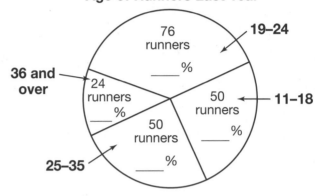

Ask Yourself
How can I find the percents represented by each age group in the graph?

Answer _____

Interpret Why is a circle graph a good way to show this data?

_____

_____

Ask Yourself
How many sets of data am I comparing?

⑤ Tiffany and Victor are comparing how far they hike. The table shows how far Tiffany hiked. The graph shows how far Victor hiked. Who hiked farther? How much farther?

Month	Tiffany
May	6 mi
June	10 mi
July	7 mi
August	10 mi

Hint Remember to make a key for your graph.

Answer _____

Determine Why is a circle graph not a good way to show this data?

_____

_____

184 Unit 6 **Using Data and Graphing**

Practice

Solve the problems. Show your work.

6 Sondra and Vin keep track of the time they spend swimming, biking, and running during a week. Sondra spent 60 minutes swimming, 70 minutes biking, and 110 minutes running. The bar graph shows Vin's results for one week. Which activity had the greatest difference in time between Sondra and Vin?

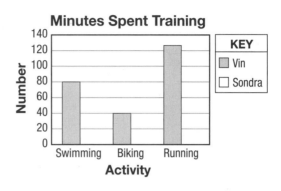

Answer _____

Consider What is another question you could ask about the problem?

_____

_____

7 George surveyed 50 students in his grade to see what game they want to play at field day. The graph shows their responses. 28% chose kickball, 26% chose soccer, 22% chose baseball, and the rest chose football. What was the most popular choice? How many students chose that game?

Answer _____

Decide Tonya says that 28 students want to play kickball. What mistake might she have made?

_____

_____

Write a problem that can be solved by making and using either a circle graph or a bar graph. Solve your problem.

Create

Lesson 22 **Strategy Focus: Make a Graph** 185

Lesson 23

Strategy Focus
Make an Organized List

MATH FOCUS: Frequency Tables

Learn

☐ Read the Problem

A park ranger needs to buy some new picnic tables. He can order small tables that seat up to 6 people, medium tables that seat up to 8 people, or large tables that seat up to 12 people. He can only order one type of table. He keeps track of the size of each group of people who picnic at the park each day. The record he made yesterday is shown at the right. If this record is like the data he collects each day, which type of table should he buy?

Number in Each Group				
8	11	4	8	8
9	6	8	3	7
4	7	7	8	5

Reread Ask yourself questions as you read the problem again.

- What kind of data is the park ranger keeping track of?
  _____

- How are the data given?
  _____

- What am I asked to answer?
  _____

☐ Search for Information

Read the problem again. Circle the information about the smallest and largest groups.

Record Write what you know about the groups.

The smallest group has _____ members.

The largest group has _____ members.

There were _____ groups that picnicked in the park yesterday.

Think about how you can use this information to solve the problem.

Decide What to Do

You know how many people each picnic table holds. You also know the number of groups and the size of each group.

Ask How can I find which type of table the ranger should buy?

- I can use the strategy *Make an Organized List*.
- I can make a frequency table to show how often each group size occurs. Then I can compare the number of groups in each size to see which type of picnic table makes the most sense to buy.

Use Your Ideas

Step 1 Find the range of group sizes.

The group sizes range from 3 to _____ people.

Step 2 Determine the intervals for the frequency table. Since the tables hold 6, 8, or 12 people, use 7–8 as an interval for a group that would sit at a medium table.

The interval for a group that would sit at a small table would be from _____ to _____ .

The interval for a group that would sit at a large table would be from _____ to _____ people.

The intervals for your frequency table do not have to be equal.

Step 3 Use the data to make a frequency table with the intervals you found.

Size of Group	Tally	Frequency
3–6	|||||	
7–8		
9–11	||	

The most frequent group size is _____ .

So the park ranger should buy _____ tables because they hold up to 8 people.

Review Your Work

Check that you tallied all the groups in the frequency table.

Recognize What is another question you could ask about the data?

_____

_____

Try

Solve the problem.

(1) A botanist at a park measured 8 trees that were between 15 and 19 feet tall. The number of trees more than 19 feet tall is one more than one fourth the number of trees that were between 15 and 19 feet tall. The number of trees less than 15 feet tall is one more than half the number of trees between 15 and 19 feet tall. How many trees did the botanist measure?

Mark the Text

■ Read the Problem and Search for Information

Identify what you know about the tree heights. Think about what you need to find.

■ Decide What to Do and Use Your Ideas

You know that the data are the heights of trees. You also know how the data for each height range are related. You must find the total number of trees the botanist measured.

Ask Yourself

How are the trees grouped?

Step 1 Make a frequency table to organize the information. Write the number of trees that are 15–19 feet tall in the table.

Heights (ft)	Frequency
Less than 15	
15–19	
More than 19	

Step 2 Find the number of trees less than 15 feet tall and the number more than 19 feet tall.

One more than half of 8 is _____ .

One more than fourth of 8 is _____ .

8 + _____ + _____ = _____ trees

So the botanist measured _____ trees in the park.

■ Review Your Work

Check that the number of each size of tree is correct.

Decide Would the labels on your frequency table change if the botanist had measured a different number of trees in each height range? Why or why not?

_____

_____

188 Unit 6 **Using Data and Graphing**

Apply

Solve the problems.

2 Mr. Long surveyed runners at a local park to find how many miles each of them ran last week. He recorded the distances for 21 runners. How many runners ran 31 miles or less last week?

Runner Distances (miles)		
27	24	28
20	21	26
25	26	28
35	28	20
35	38	26
35	26	21
36	43	20

Distance (miles)	Tally	Frequency
20–25		
___–31		
38–43		

Hint Use the range of distances to find intervals for making a frequency table.

Ask Yourself
Do I need to know how many runners were surveyed?

Answer _____

Explain How did you complete the *Distance* column in the table?

_____

3 A community band gives free weekly concerts in the park. Six band members are aged 40–49. Two more are aged 20–29 than are aged 40–49. Twice as many are aged 20–29 as are aged 30–39. The rest are aged 50–59. If there are 23 band members in all, which age group has the fewest members?

Ages	Frequency
20–29	
30–39	
40–49	
50–59	

Ask Yourself
Which age group should I find first?

Hint Find out how many people are in each age group.

Answer _____

Conclude How did making a frequency table help you solve this problem?

_____

Lesson 23 **Strategy Focus: Make an Organized List** 189

Ask Yourself

How old is the youngest person who comes to the park at least once a week? The oldest person?

Hint The age ranges cannot overlap in your frequency table.

4) A community group surveyed fifty young people to find out if they come to the park at least once a week. The age of each person who said yes was recorded. Which age group had the most people who came to the park at least once a week?

Ages of Park Visitors			
10	6	8	18
14	17	13	12
15	14	7	19
17	11	15	11
9	12	19	12
14	8	12	11

Age Range	Tally	Frequency
5–9		
___–14		
___–19		

Answer _____

Clarify Why are only 24 numbers given for the data instead of 50?

_____

_____

Hint Use the data to find the number of campers in each age group.

5) Julia thinks the largest group of campers is aged 8 to 10. Mark thinks the largest group is aged 11 to 13. There are nine 14–16 year-olds. There are $\frac{2}{3}$ as many 5–7 year olds as 14–16 year olds. There are twice as many 8–10 year olds as 5–7 year olds. If there are 40 campers in all, who is correct? Explain.

Age Range	Number
5–7	
11–13	

Ask Yourself

What operations can I use to find the number of campers for each age group?

Answer _____

Apply How did you find the number of 11–13 year old campers?

_____

_____

Practice

Solve the problems. Show your work.

6 Skateboarders practice in the park for a skateboard contest in the summer. Of 14 skateboarders, 4 practice 1–2 hours a day, 6 practice 3–4 hours a day, 2 practice 5–6 hours a day, and the rest practice 7–8 hours a day. How many skateboarders practice more than 4 hours a day?

Answer _____

Determine Could you answer the question if the total number of skateboarders was not given?

_____

_____

7 A few community groups make ice sculptures in their park during the winter. A park official recorded the lengths of the sculptures that were made last year. How many more ice sculptures had lengths of 6–10 feet than had lengths of 11–15 feet?

Ice Sculpture Lengths (ft)					
8	16	9	12	17	18
10	8	16	7	6	20

Answer _____

Interpret How many ranges of length did you use to solve the problem?

_____

_____

Create

Write a problem that involves ranges of data and that can be solved using the strategy *Make an Organized List*. Solve your problem.

Lesson 24

Strategy Focus
Make a Graph

MATH FOCUS: Line Graphs

Learn

▪ Read the Problem

The table shows the number of endangered animals and plants from 1980 to 2005. Between which two years did the number of endangered plant species become greater than the number of endangered animal species? In which recorded year was the difference between the number of endangered plants and the number of endangered animals the greatest?

Year	Animals	Plants
1980	174	50
1985	207	93
1990	263	179
1995	324	432
2000	368	593
2005	389	599

Reread Ask yourself questions as you read.

- What is the problem about?

  _____

- How is the data presented?

  _____

- What do I need to find?

  _____

▪ Search for Information

Review the information in the table.

Record Write what you know about the data.

The least and greatest numbers of endangered animal species are _____ and _____ .

The least and greatest numbers of endangered plant species are _____ and _____ .

The time period is from _____ to _____ .

Think about how you can use this data to solve the problem.

192   Unit 6   **Using Data and Graphing**

Decide What to Do

You know the number of endangered animal and plant species from 1980 to 2005.

Ask How can I compare the two sets of data?

- I can use the strategy *Make a Graph*. I can make a double line graph.
- I can use the graph to see when the number of endangered plants became greater than the number of endangered animals. I can also use the graph to see which year had the greatest difference between the numbers for animals and plants.

> Double line graphs are useful for comparing two sets of data over time.

Use Your Ideas

Step 1 Draw and label the axes for your graph. Use intervals of 5 years. For the vertical axis choose a scale with equal intervals. The least number is _____. The greatest number is _____. Complete the graph.

Step 2 Look at the graph. Use it to answer the questions.

Endangered Animal and Plant Species

KEY
— Animals
--●-- Plants

The number of endangered plants became greater than the number of endangered animals between the years _____ and _____.

The difference between the numbers of endangered plant and animal species was greatest in the year _____.

Review Your Work

Check that all available data is included in your graph.

Generalize How did the double line graph help you solve the problem?

Try

Solve the problem.

(1) Antonio planted bean seeds in March and recorded the growth of one plant for five months. Predict how tall the bean plant will be in September.

Month	March	April	May	June	July	August
Height (in.)	0	5	15	30	37	38

Read the Problem and Search for Information

Reread the problem. Identify the two types of measurements in the table.

Decide What to Do and Use Your Ideas

You can use the strategy *Make a Graph* to show the data in the table.

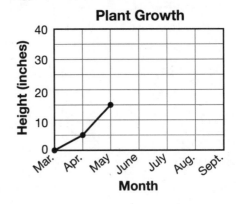

Step 1 Draw and label the axes for your graph. Label the months on the horizontal axis and choose a scale for the vertical axis.

Does the height increase or decrease?

Step 2 Complete the graph and look for a trend. A trend shows you how a graph changes. The height is increasing more slowly at the end.

So the bean plant should be about _____ inches tall in September.

Review Your Work

Check that you plotted all the data shown in the table.

Describe How did you use the graph to solve the problem?

_____

_____

194 Unit 6 Using Data and Graphing

Apply

Solve the problems.

2 Dr. Wu tracks the distance traveled by a female hyena over several hours. The results are shown in the table. About how far would you expect the hyena to travel in 7 hours?

> **Hint** Make a line graph to record the hours and distance.

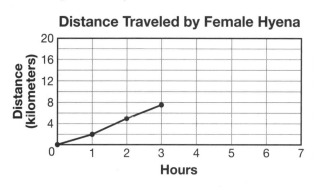

Time (hr)	Distance (km)
0	0
1	2
2	5
3	7.5
4	10.5
5	13
6	15.5

> **Ask Yourself**
> Does the graph show a trend?

Answer _____

Conclude Can you find the exact answer? Why or why not?

_____

_____

3 The number of birds sighted is sometimes listed as number per hour. The numbers of bobwhites and herons seen per hour are shown in the table. Are the numbers changing in the same way? Explain.

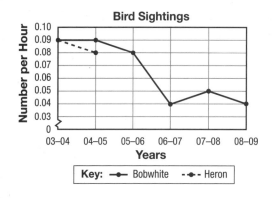

Year	Bobwhite	Heron
2003–04	0.09	0.09
2004–05	0.09	0.08
2005–06	0.08	0.07
2006–07	0.04	0.05
2007–08	0.05	0.06
2008–09	0.04	0.07

> **Hint** You can plot the data on a double line graph.

> **Ask Yourself**
> Will the broken vertical scale on the graph affect my work?

Answer _____

Explain How did the completed graph help you find the answer?

_____

_____

Lesson 24 **Strategy Focus: Make a Graph** 195

Ask Yourself

Does the number of eggs in each clutch increase or decrease at a similar rate?

Hint A prediction is an estimate.

④ Ashley has two adult bearded dragon lizards. The female bearded dragon lizard lays clutches of eggs. The number of eggs she lays in each clutch is shown in the table. Predict how many eggs the dragon will lay in her next clutch.

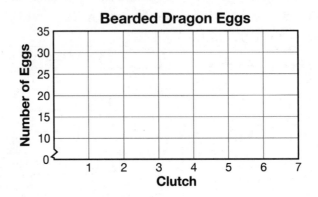

Clutch	Number of Eggs
1	16
2	24
3	26
4	24
5	28
6	31

Answer _____

Examine What information is given that is *not* needed?

_____

_____

⑤ Pedro and Kia grow tomatoes. The table shows the pounds of tomatoes they harvested over several seasons. Predict how many more pounds Pedro will grow than Kia in Year 7.

Year	Pedro	Kia
1	18	15
2	21	23
3	30	30
4	32	24
5	33	20
6	35	16

Ask Yourself

Are the weights of Pedro's tomatoes always greater than Kia's tomatoes?

Hint Remember to use a key for your graph.

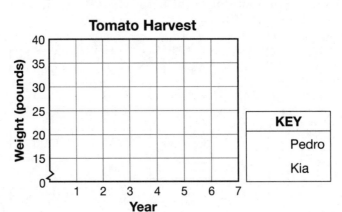

Answer _____

Determine Explain how you used the graph to answer the question.

_____

_____

196 Unit 6 Using Data and Graphing

Practice

Solve the problems. Show your work.

6) Jenna kept track of the growth of an Alaskan pea plant over 50 days. About how tall was the plant at 40 days?

Day	10	20	30	35	50
Height (cm)	0	10	40	50	90

Height of Alaskan Pea Plant

Answer _____

Sequence What steps did you take to solve this problem?

_____

_____

7) The graph shows the number of farms in Nebraska over a 25-year period. The table shows the number of farms in North Dakota over the same period. In which year was there the least difference between the number of farms in North Dakota and Nebraska?

Year	Farms in North Dakota
1980	40,000
1990	34,000
2000	30,000
2004	30,000

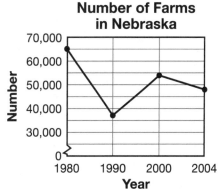

Number of Farms in Nebraska

Answer _____

Formulate What is another question you could ask?

_____

_____

Create

Write a new problem that can be solved by using the strategy *Make a Graph*. Then solve it.

Lesson 24 **Strategy Focus: Make a Graph** 197

UNIT 6 Review

In this unit, you worked with three problem-solving strategies. You can often use more than one strategy to solve a problem. So if a strategy does not seem to be working, try a different one.

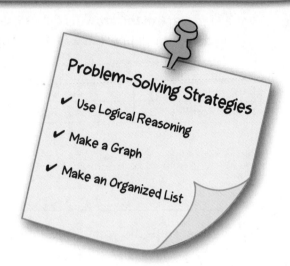

Problem-Solving Strategies
- ✔ Use Logical Reasoning
- ✔ Make a Graph
- ✔ Make an Organized List

Solve each problem. Show your work. Record the strategy you use.

1. Last Saturday, Jason and four of his friends went fishing at a nearby lake. He started to make a bar graph to show how many fish he and his friends caught. Cathy caught 12 fish. Luna caught 10 fish. Juan caught 8 fish. Who caught about twice as many fish as Juan?

   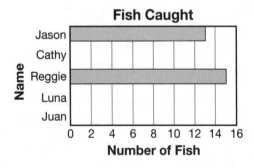

   Answer _____

   Strategy _____

2. A park ranger counted the number of moose he saw in a swamp each day for 8 days. On those days, he saw 14, 20, 17, 32, 17, 21, 12, and 19 moose. What are the median and the mean of the data?

   Answer _____

   Strategy _____

198   Unit 6 **Using Data and Graphing**

3. Asha planted a flower in front of her house. She measured the height of the flower as it grew. About how tall was the flower at 5 weeks?

Weeks	1	2	3	6
Height (in.)	0.5	2.5	5	8.5

Answer _____

Strategy _____

4. The data in the stem-and-leaf plot below include an outlier. How does removing the outlier affect the mean, median, and mode of the data?

Stem	Leaf
0	5 8 9 9
1	3 4 6 9
2	3 6 8
3	3 9
4	
5	
6	1

Key: 6 | 1 = 61

Answer _____

Strategy _____

5. A weather forecaster is comparing the average monthly temperatures for Caribou, Maine and Portland, Maine. The line graph shows the temperatures for Caribou. The table shows the temperatures for Portland. In which month was the difference in temperatures between Caribou and Portland the least?

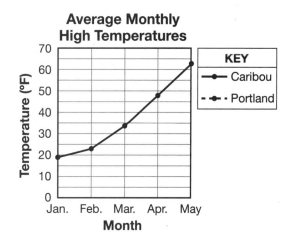

Portland, Maine

Month	High(°F)
January	31
February	34
March	42
April	53
May	61

Answer _____

Strategy _____

Explain how you found your answer.

Solve each problem. Show your work. Record the strategy you use.

6. Mrs. Reid polled 150 students about their favorite type of book. The circle graph shows their responses. How many students chose science fiction?

Favorite Types of Books
Science Fiction
Adventure 32%
Fantasy 20%
Mystery 28%

Answer _____

Strategy _____

7. Chad thinks that the top box plot shows data with a greater range. Is he correct? Explain why or why not.

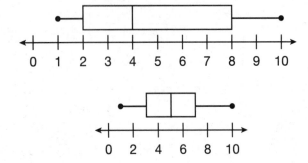

Answer _____

Strategy _____

8. Jed and Ling started a bar graph to show the animals they saw. Ling counted 15 squirrels, 10 rabbits, 5 deer, and 8 chipmunks. For which animal was there the greatest difference between the numbers that Jed and Ling counted?

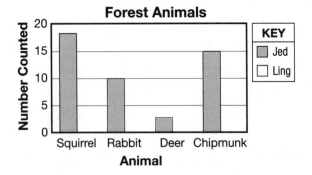

Answer _____

Strategy _____

Explain why a double bar graph is a good way to show this data.

9. The ages of people in a group visiting a historic site are 77, 53, 71, 29, 47, 31, 44, 61, 1, 20, 65, and 80. How many more people that are older than 60 are in the group than people that are younger than 20?

Answer _____

Strategy _____

10. A county has 120 parks. Of those, 75 are less than 2 acres. Two-thirds of the other parks are less than 4 acres, but at least 2 acres. How many more parks are there that are at least 2, but less than 4 acres than parks that are at least 4 acres?

Answer _____

Strategy _____

Write About It

Look back at Problem 10. Describe how you used the information in the problem to choose a strategy for solving the problem.

_____

_____

_____

Team Project: Survey Your Class

Your team is planning a trip for the whole class. Together with your teammates, find what outdoor activities students would like to do.

Plan
1. Work together to make a list of 5 outdoor activities.
2. Take a survey. Ask the rest of the students in your class which activities are their favorites.
3. Record which is the favorite of each person that you survey and whether the person is a boy or a girl.

Decide As a team, choose the type of graph that you will use to display the results of your survey.

Create Make a graph that you will use to share the results.

Present As a team, share your results with the class.

Unit 6 Review

Glossary

A

adapt make changes to suit a new situation

analyze study carefully and come to conclusions

apply put to use

area a measure of the amount of surface enclosed in a figure, usually given by the number of square units needed to cover the figure

assess study carefully and determine the quality of

B

bar diagram a diagram that organizes information and helps determine calculations to solve a problem

C

categorize put in a group

central angle an angle whose vertex is at the center of a circle

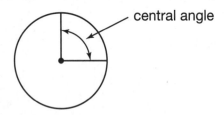

circle graph a graph in the shape of a circle that shows parts of a whole

clarify make something easier to understand

conclude make a decision after careful thinking and reasoning

congruent having the exact same size and shape

consider think carefully about

contrast compare the differences

cross multiply to verify or solve proportions by multiplying the numerators by the opposite denominators and setting the results equal

cubic foot a unit used to measure volume

D

data a set of information, usually numbers

decide make up your mind

decimal a number containing a decimal point

demonstrate show clearly

denominator the number below the bar in a fraction

depth the distance from the front to back of a 3-dimensional figure

determine come to a conclusion

develop create and grow

diameter a line segment that passes through the center of a circle and whose endpoints are two points on the circle

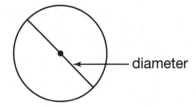

difference the result of subtraction

differentiate identify differences

discount a decrease in price

distinguish recognize by differences between

dividend the number that is divided in a division problem

divisor the number by which another number is being divided

double bar graph a graph that uses bars to compare two sets of data

double line graph a graph that uses line segments to compare two sets of data

E

edge a line segment where two faces of a solid figure meet

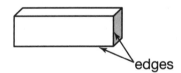

employ use

equation a mathematical statement that two expressions have the same value

equivalent ratios ratios that can be represented by equivalent fractions

evaluate examine the value of

expression a mathematical phrase containing numbers, variables, and/or operations with no equals sign

extend go further; continue

F

face a flat surface of a solid figure

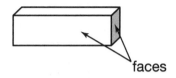

favorable outcome an outcome that is part of the event that you want to find the probability of

formulate create

frequency table a table for organizing data showing how often each type of data occurs

function table a table in which each input value has one output value

G

generalize make a general statement

I

infer come to a conclusion from reasoning

integer a number that is a whole number or the opposite of a whole number

interpret explain the meaning of

inverse operation an operation that reverses the effect of another operation

Example: Addition and subtraction are inverse operations.

investigate discover and examine facts

J

justify show proof

L

length the distance from one end of a line segment or curve to the other

M

mean the sum of a set of numbers divided by the number of numbers in the set

measure of central tendency a measure that is used to describe a set of data

Example: The mean, median, and mode of a data set are all measures of central tendency.

median the middle number in a list of numbers arranged in order, or if a set contains an even number of numbers, the mean of the two middle numbers

mixed number a number that is greater than 1 that is written as a whole number and a fraction

Example: $2\frac{1}{2}$

mode the number (or numbers) that occurs most often in a set of data

modify change

multiple A multiple of a number is the product of that number and some whole number greater than 0.

multiply find the product of numbers

N

numerator the number above the bar in a fraction

O

outcome a possible result in a probability experiment

outlier a number that is much greater or much less than most of the other numbers in a data set

P

pattern objects or numbers arranged according to a rule or rules

percent a ratio of a number to 100 using a percent sign

perimeter the distance around an object

prime number a whole number greater than 1 whose only two factors are 1 and itself

probability the likelihood that an event occurs

product the result of multiplication

proportion an equation stating that two ratios are equal

Pythagorean theorem In a right triangle, the sum of the squares of the lengths of the two legs is equal to the square of the length of the hypotenuse.

Q

quartile one of the numbers that separate a set of data into four equal parts

R

radius a line segment that connects the center of a circle to a point on the circle

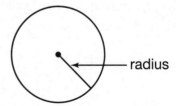

rate a ratio that compares quantities measured in different units

Example: $\frac{50 \text{ miles}}{\text{hour}}$

ratio a comparison by division of two quantities

Example: 5 to 3

reciprocal The reciprocal of a number is a second number such that the product of the two numbers is 1.

Example: The reciprocal of $\frac{3}{1}$ is $\frac{4}{3}$.

recognize identify and understand

rectangular prism a solid figure with two parallel, congruent, rectangular faces called bases and four other rectangles for faces

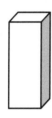

S

scale an arrangement of numbers in order at uniform intervals labeling the axes of a graph

sequence put in order

square foot a unit that is used to measure area

support give good reasons

T

tree diagram a diagram with branches that represent possible combinations of outcomes

V

variable a letter or symbol that represents a quantity that can change

Venn diagram a diagram that uses overlapping circles to organize and show data

volume a measure of the amount of space enclosed by a solid figure

W

width the measure of one side of a 2-dimensional or 3-dimensional figure